T0077528

Order this book online at www.trafford.com
or email orders@trafford.com

Most Trafford titles are also available at major online book retailers.

Print information available on the last page.

ISBN: 978-1-4120-9606-5 (sc)

Trafford rev. 03/15/2021

Trafford PUBLISHING www.trafford.com
North America & international
toll-free: 844-688-6899 (USA & Canada)
fax: 812 355 4082

AERIAL VIEW, TERERRO, NEW MEXICO, AUGUST 1931

TERERRO

✲✲✲✲✲✲✲✲✲✲✲✲✲✲✲✲✲✲✲✲✲✲✲✲✲✲✲✲

by **Leon McDuff**

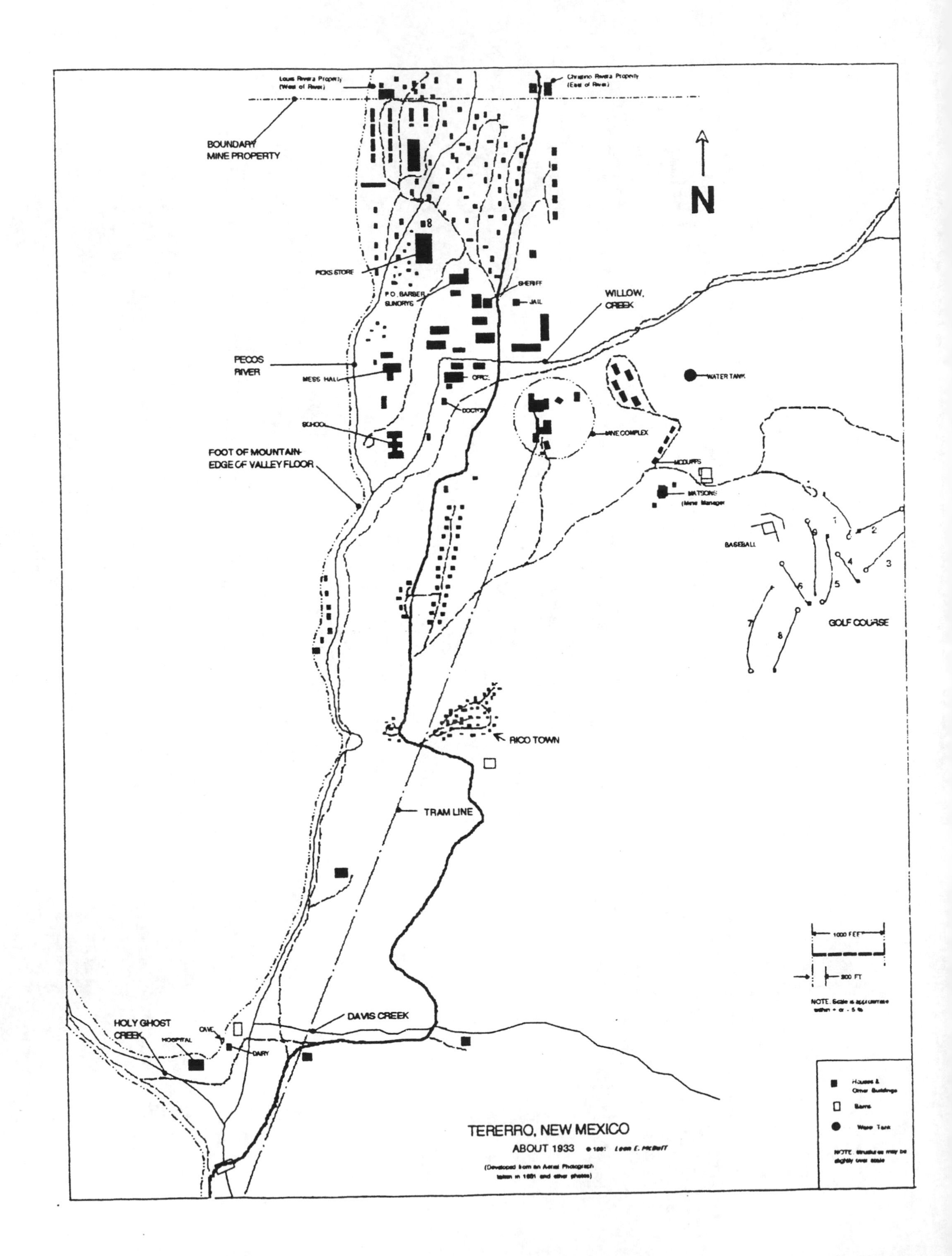

Louis Rivera Property
(West of River)
Christino Rivera Property
(East of River)
BOUNDARY
MINE PROPERTY
N
PICKS STORE
P.O. BARBER
SUNDRYS
SHERIFF
JAIL
WILLOW
CREEK
PECOS
RIVER
MESS HALL
WATER TANK
SCHOOL
DOCTOR
MINE COMPLEX
FOOT OF MOUNTAIN-
EDGE OF VALLEY FLOOR
MATSONS
(Mine Manager)
BASEBALL
1
2
3
4
5
6
7
8
9
GOLF COURSE
RICO TOWN
TRAM LINE
1000 FEET
200 FT
NOTE: Scale is approximate
within + or - 5 %
HOLY GHOST
CREEK
CAVE
HOSPITAL
DAIRY
DAVIS CREEK
TERERRO, NEW MEXICO
ABOUT 1933
Houses &
Other Buildings
Barns
NOTE: structures may be
slightly over scale

✿ ✿ ✿ ✿ ✿ ✿✿ ✿

TERERRO

✿ ✿ ✿ ✿ ✿ ✿ ✿ ✿

by

Leon E. McDuff

The history of a New Mexico mining town
called

TERERRO

and the boyhoods of two who lived t here

First printing, 1993
First revision, 1996
Second revision, 1999
Fourth revision, 2006

Library of Congress Catalog Card Number 93-91654

Published by author
107 Scot Court
Green Valley, CA 94534

DEDICATED

TO MY DEAR WIFE

Lois

whose loving patience, guidance,
and persistence in demanding the best from me
is reflected in these words,

✲ ✲ ✲ ✲ ✲ ✲ ✲

AND TO THE MEMORY OF
MY BELOVED FRIEND OF 60 YEARS

Joseph Tousley Matson, Jr.

whose friendship as a boy
made life so enjoyable.

ACKNOWLEDGEMENTS

It is a pleasure to thank a great many people who have helped me put this book together.

I owe a deep debt of gratitude to Ken Paulsen, Vice President of AMAX Resource Conservation Company, for first suggesting and then allowing me total access to their files on the American Metal Company of New Mexico. Further, for providing me with the latest information relative to the environmental problems at the mine and mill sites. Also to Kathryn Seprino, Manager of AMAX's Information Services for her assistance in recovering the needed files from storage. Without their help, this book would have been far less detailed and, in some instances, far less accurate.

I am especially grateful to Lew Howes, my dear friend, for encouraging me to write the book and for his help in copying the several hundred documents during my research at AMAX's offices.

I want to thank May Walter, my 5th and 6th grade teacher, for her encouragement, for providing me with pictures taken by her husband Harold, and for reminding me that she had taught me how to spell better than I had in some of the early chapters I sent to her for comment.

I am indebted to the following school classmates, and to other former residents of Tererro or Alamitos who have provided short anecdotes about life in Tererro and at the mill: Maxine Smith Felton, Dan Smith, Norman Littrell, Wayne Niemon, Herbert Pritz, Ralph Littrell, Jr., Larry Earickson, Frances Earickson Slade, Dorothy Barker McNeese, and Lee Kennedy. Also to Huie and Sherry Ley, current residents of Tererro.

To Bonai and Cecil Sanders, Jan Greene, Mary and Don Montgomery, my brother, Delta and his wife, Dodie, and my sister, LaVoyse, I am thankful for their having read various and sundry chapters and encouraged me to continue writing.

To the family of my dear friend Joe, to whom this book is dedicated, I owe a special debt of gratitude for their enthusiastic encouragement—to Lilah, his widow, to his daughters, Margo, Ceil, Ann, Sarah, and Kate, to his son, Bob, to his sister, Isabel, and to his niece, Patricia.

To my sons, Russell, Kevin, and Kenneth, my daughter, Robin and her life companion, Leslie, my daughters-in-law, Tobae, Shelley, and Karen for their encouragement.

Finally one special debt—to my loving, patient wife, Lois, for her encouragement and support, for her proof-reading and editing, and for insisting that I not be satisfied with anything but the best as the final product of my endeavors.

CONTENTS

✲ ✲

PREFACE

✲ ✲

Spawned by snow-clad mountain peaks, dozens of turbulent creeks and streams flow from the Sangre de Cristo Mountains located in north-central New Mexico. They flow swiftly to the Pecos River below whose source lies deep within the almost quarter-million acres of the Pecos Wilderness. Within the wilderness boundaries, the peaks rise majestically above the 12,000 foot timberline; below are dense forests interspersed with alpine meadows where wildflowers bloom profusely during July and August. There are also numerous natural lakes and many colorfully named Pecos River tributaries—Holy Ghost Creek, Panchuela Creek, Rio Mora, Rio Valdez, Rito de los Chimayosos, Rito las Trampas, and Noisy Brook to name but a few.

Two dozen miles to the south, this range of the Rocky Mountains gives way to the high Glorieta Mesa. At the small, picturesque town of Pecos (founded by Franciscan friars sometime in the 1600's) the river emerges from the narrow Pecos River canyon extending about 35 miles to the north. At Pecos, the canyon broadens abruptly, and the river flows gently to the Rio Grande on the Texas/Mexico border some 450 miles further south.

The broader river flatlands provide relatively fertile land on which to grow crops during the short growing season—squash, maize, frijoles, chilies. A Franciscan mission, located for almost two centuries at a nearby Indian pueblo, relied upon these farmlands for much of its food.

Little remains of the Mission and the once thriving pueblo of around two thousand habitants. There are the stone foundations and deteriorating adobe walls of two mission churches (the first built in the 1620's, the other in the early 1700's). Also, there are the excavated ruins of the pueblo itself. Administered by the National Park Service, Pecos National Historical Park is now its name. Abandoned in 1838, Cicuye was originally its name.

Fourteen miles to the north, the canyon is generally quite narrow and the mountains rise steeply on either side. Where Willow Creek joins the Pecos, there was once another community of over two thousand habitants. There are few remains of this once bustling and more modern, yet somewhat primitive, community. There are a few deteriorating concrete foundations, some deeply rutted abandoned roadways, rusted parts of 1920's and 1930's automobiles, and a large pile of gray-green waste rock or gangue. Administered by the New Mexico Department of Game and Fish, Willow Creek Campground is now its name. Abandoned in 1939, Tererro was then its name.

Of the many who visit Pecos National Historical Park each year—to look and to learn—none depart without knowing something of its history. Of the many who visit Willow Creek Campground each year—to fish, to hunt, to picnic, to relax—not many depart with any knowledge of its history. Most do not realize that they may have picnicked on a spot where once stood a store, a school, a post office, a sheriff's office or a jail. Perhaps it was a tent house, a tar-papered shack, a mess hall, an office, a bunkhouse, or one of the many hundreds of outhouses which once defiled the beauty of this place. This was

Tererro, the site of a large zinc, lead, copper, silver and gold mining operation. At the time, it had the largest payroll in the state of New Mexico. The name, colloquially translated from Spanish, was thought by many early residents to mean *deerlick*, and it was chosen for that reason; it was not chosen because it means *mine dump* as conjectured by some who never lived there and who have written short articles about its history.

Downriver, near the confluence of the Pecos and Holy Ghost Creek, there is another place now called Tererro. It consists of a couple of houses, a general store, a post office, and some horse stables. The owners and proprietors are Huie and Sherry Ley. In addition to the store, they provide horses and guides for rides or pack trips into the Pecos Wilderness just a few miles away.

This is not the Tererro of old; the two have only two things in common—the names and the locations. The American Metal Company of New Mexico gave birth to the old Tererro when it launched the mining operation near the northern boundary of its property in December, 1925. The southern boundary of its property, two miles south, lies near the new Tererro. According to the U. S. Post Office, the spelling of the new is the same as that of the old (*Tererro*). According to most maps and the State of New Mexico, the pronunciation of the new is the same but the spelling is different (*Terrero*).

By the end of this century, there may be none living who can relate Tererro's history from first-hand experience. Even now, few can recount Tererro's 1925 beginning through its 1939 untimely demise. The author, having lived there longer than all but two other individuals, is one who still can. For him, it holds fond memories—memories recorded herein for his descendants, and for his friends who lived there, and for their descendants. The posterity of those who lived there can probably never experience the same type of primitive living nor have the opportunity to live in such an interesting and enchanted place.

The town would never have existed were it not for the mine, and its history has almost faded into antiquity. There have been magazine articles, mining and geologic reports, a graduate student's thesis and other sundry articles which have touched on that history, but none have covered it in any detail. Some have provided inaccurate information.

All save one of the principal participants in the mine's history are now deceased. Also recorded herein, therefore, is a chronicle of that operation condensed from hundreds of letters and reports written by those principals, letters and reports which normally would have been destroyed years ago, but which were unaccountably saved from destruction. Well over three-fourths has been taken from "*Narrative Reports*" written by the mine manager every ten days during the first two years, and every two months thereafter.

I have chosen to divide this work into two parts—one to chronicle the history of the mine, and the other to chronicle the history of the town. The two, however, are inseparable, for neither history is complete without the other.

Nature has almost reclaimed the place that was once Tererro. Hopefully, through this book, it will live again, if only in someone's imagination. This book is the legacy of one who lived there and loved it.

IN THE BEGINNING

* *

Roughly 200 million years ago, at the beginning of the Mesozoic era, the sea dominated the landscape where now lie the majestic Rocky Mountains of the North American Continent. The few areas of land rising above the sea were relatively flat. Thick sedimentary rock beds, deposited and compressed by the seas over thousands of millenia, made up most of the visible dry land vista. Scattered volcanic cones also spewed molten rock and ashes. Mixed within the sedimentary beds were bodies of dark igneous rocks deposited over the ages by intrusive volcanic flows of eons gone by. Some areas, eventually covered by thick beds of decaying vegetation, would subside below the seas again, there to be covered by more sediments and compressed into thick beds of coal or pockets of crude oil.

Earth's crust consists of several great masses known as tectonic plates. They are in constant, slow, relentless motion as they float on the earth's liquid center. As one plate collides with another, the weaker of the two is slowly and inexorably compressed. Beginning at the end of the Palezoic era, and continuing over the Mesozoic era of perhaps 130 million years, the compression caused by the plate movements resulted in tight crumpling, folding and uplifting of the sedimentary rock beds to heights some two to three miles above sea level. Fracturing and faulting occurred when the stresses became too great. Both horizontal and vertical movement of the rock occurred along the fault lines and fissures. In the process, pockets and voids were also created.

Magma from deep within the earth's bowels welled up to fill these pockets and voids. Hot, silica-bearing solutions in the magma formed quartz veins as they cooled. In a process known as mineralization, the solutions often contained rich concentrations of metallic ores—occasionally mixtures of lead, copper, and zinc compounds. Rarely, they might also contain silver and gold. In addition, both the sedimentary and the igneous rock were often recrystalized by the heat and pressure of the compressive forces into dense, metamorphic equivalents such as micaceous diorite, schist, and quartzsite.

The forces of nature—wind, rain, glaciers, earthquakes—immediately began to erode the uplifted terrain, sculpturing lofty peaks and deep valleys as we now know them. The sculpting continues even today, and the residue moves doggedly toward the sea, carried by glaciers and thousands of rivers and streams on their journey back to the seas whence came the water.

Thus, in the simplest of terms, the Rocky Mountains came to be—a geologic mystery; a mystery that man would unravel in less time than that required for a mere fraction-of-an-inch of sedimentary rock to be formed.

During the approximately 70 million years of the Cenozoic era, some of the ore bodies deposited within the conglomerate mass of the mountains were exposed by the wearing down and erosion—some probably eroded away in their entirety. One such large ore body, containing a high concentration of lead, zinc, iron, and copper ores, including some native copper, was ultimately exposed. That outcropping, approximately 1,000 feet long, lay on the northern slope of Willow Creek Canyon near its juncture with the Pecos River Canyon of North Central New Mexico. It first saw light of day many thousands of years before Columbus discovered America. There it would remain undiscovered by Anglos for almost four hundred years after America's discovery. It is possible that, much earlier, Native Americans found the outcropping and made use of the native copper.

PECOS MINE, 1930

part 1

✿ ✿

The Mine

✿ *its derivation*
✿ ✿ ✿ *its duration*
✿ ✿ ✿ ✿ *its demise*

chapters
1 thru 15

chapter 1

* *

Discovery and Development

In 1872, President Grant, in a move to encourage settlement of the West, proposed and Congress passed the Mining Act of 1872, or the "Apex Law," that is still in effect as of this writing. Currently, there is a move in Congress to change the law. That law allows miners and mining companies free access to federal land. Even today, they may file claim to, or patent, those lands at the 1872 cost of $2.50 per placer-claim acre and $5.00 per lode-claim acre. There are, of course, numerous provisions of the law, as well as provisions of individual state laws, that prescribe those things required before a patent can be filed.

Seven years later, in 1879, the Santa Fe Railroad opened rail service along the old Santa Fe Trail to Santa Fe and the New Mexico Territory. The railroad provided prospectors easy access to the mountains of New Mexico to stake potential mining fortune claims. One such prospector, J. J. Case from Kansas, boarded a Santa Fe train and headed for the Sangre de Cristo range of the Rocky Mountains east of Santa Fe. The year was 1881.

He ended up prospecting on the upper Pecos River around Willow Creek, an elevation of 8,000 to 8,200 feet, and in 1882, discovered the 1,000-foot long out-cropping. The copper caught his fancy; he wasn't interested in the zinc and the lead, nor was he aware that there were sizable quantities of gold and silver also present in the ore-body. Thereupon, he developed and identified the claim to the extent required by law and filed for a patent. He called it the Evangeline—perhaps the name of his mother, or sister, or an old sweetheart. He hired a couple of helpers; he named his endeavor the Pecos River Mining Company.

On June 3, 1881, another event occurred that was to have a direct bearing on Case's discovery—Joseph Tousley Matson was born in Johnson City, Tennessee. Ultimately, he would be the one responsible for mining and milling the ore which Case had set out to discover in the year of Matson's birth.

The outcropping Case discovered was about 300 feet up the mountainside from Willow Creek. About 160 feet lower, Case opened up an adit into the mountain to a point directly below the outcrop, a distance of about 140 feet. This placed him within the main Evangeline ore body. From the outcrop above, he sank a 180-foot-deep shaft to intersect and extend below the adit. This development work he called the Hamilton Mine.

Case worked diligently to develop the Evangeline ore body. At the same time, he looked for other outcroppings of ore. He found others in the same general area and filed patents on several. The most important, about 200 feet to the southwest, he called the Katy Did. Another, which didn't prove to be of value, he called the Katy Didn't.

Not long after, A. H. Cowles and his brother,[1] whose first name or initials have long since been forgotten, were shown the Hamilton mine. They convinced their father to purchase the Pecos River Mining Company from Case. Included was the Hamilton Mine and several other mining claims as well. In 1886, the brothers attempted to further develop the Hamilton after changing its name to the Cowles Mine. Little effort was required to convince them the complex ore[2] could not be refined and smelted in the United States. Work was stopped and the mine lay idle for the next 17 years.

In 1903, the elder Cowles, after having heard news of some new treatment technology, decided to try again. He formed the Pecos Copper Company and tried for four years to make it a profitable venture, but to no avail. Only a small amount of "shipping-grade" copper ore was produced by tedious hand sorting. It then had to be transported by mule train over twenty-two miles of rugged mountain roads to Glorieta, the nearest station on the Santa Fe Railroad. Admitting failure, Cowles abandoned the venture and the mine lay idle for another nine years.

Benjamin Franklin Goodrich, founder of the B. F. Goodrich Rubber Company had two sons, David and Charles, who at the ages of 36 and 34 respectively, joined with Henry Lockhart in 1912 to found the Goodrich-Lockhart Company. Although the Certificate of Incorporation from the State of Delaware shows the name as the Good-Rich Lockhart Company, all other references found show the dash where it would logically be expected—between Goodrich and Lockhart. No information has been found concerning Lockhart's role in the corporation, but one can probably assume he provided financial backing and was a silent partner in the venture. David was President and Charles was Secretary.

The stated purpose of the company was to locate, lease, acquire, hold, own, mine, manage, operate, license, prospect, promote, develop, or otherwise deal in or dispose of mines, mining claims, lodes, placers, prospects, and mineral lands of every kind and description. Incorporated in the State of Delaware, the intended area of operation was in the Southwest and, more definitively, in the State of New Mexico.

The first potentially profitable property located was Cowles' Pecos Copper Company. They worked and bargained with him for four years and, in 1916, took an option to purchase the company if the mine could be developed to show a potential profit. Development was energetically pursued (except during the war years of 1917-1918) until 1921.

During that period, some 150 tons of ore were mined and stockpiled, and both the Evangeline and Katy Did ore bodies were developed to the 400-foot level. A new shaft close to Willow Creek was sunk 400 feet into the Katy Did. The Evangeline shaft was further sunk to 500 feet. Drifts were opened between the two ore bodies at the 300 and

---

[1] The little summer resort town of Cowles, about six miles north of Willow Creek at the confluence of the Pecos River and Winsor Creek was named for their father.

[2] The ore was a conglomerate mixture of galena (lead), sphalerite (zinc), chalcopyrite (copper), and pyrite (iron). The gangue (waste rock) was a sheared micaceous diorite.

400 levels as well as radially to the ore body walls. Considerable diamond drilling was done from the 400-foot level into both ore bodies that helped define their sizes and

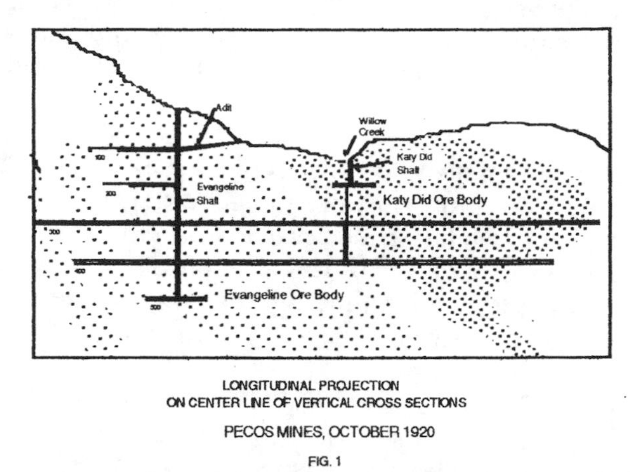

LONGITUDINAL PROJECTION
ON CENTER LINE OF VERTICAL CROSS SECTIONS

PECOS MINES, OCTOBER 1920

FIG. 1

shapes. Bore holes were made to the 1,000-foot level in the Evangeline, and to the 800-foot level in the Katy Did. See Fig 1

In November, 1920, R. M. Raymond, a professor of mining engineering, wrote a report on a study he had made on the mine for an unidentified client.[3] The study examined in detail several known methods of milling the complex ore, start-up and production costs based on the various milling methods, and the potential profit should the mine be purchased and put into production. In the study, he presented a geological report by a Mr. Basil Prescott whom he had engaged to study the geology of the area with respect to the potential depth of the ore bodies. Prescott's study indicated the ore bodies would probably extend to quite a depth, but would be irregular in shape.

Raymond also included in his report estimates of the ore reserves and concentrations as presented to him by a Mr. Leggett of Goodrich-Lockhart. Leggett had engaged Butters' Laboratory of Salt Lake City to conduct the concentration tests. Using Butters' Lab tests (conducted by a Mr. Jones), and the ore reserve figures (provided by Leggett), Raymond evaluated the cost of various options for mining, transporting, milling, and smelting the ore based on milling, smelting, and transportation technology extant at that time. He was provided with ore samples from the mine by Mr. Leggett and, although he perceived that the sampling was "carefully done," he nonetheless engaged a Mr. Lindberg from Los Angeles to double-check Leggett's sampling procedures. The final conclusion was that "...the ore bodies were limited slightly as compared to the bodies and values presented by Mr. Leggett."

Leggett's ore samples were sent to The General Engineering Company in Salt Lake City for further concentration tests, and the results indicated concentrations of the minerals at somewhat below those obtained by Jones at Butters' Laboratory. Leggett then instructed Jones to further conduct tests to determine if a clean lead-product could be obtained from the ore. The results were positive, but the process was far more complicated than previously used.

---

[3] No reference has been found as to where Raymond was a professor, or who commissioned him to do the study.

Raymond then had some of the higher grade ore hand-sorted and tested for concentration with good results, enough so that it seemed perhaps worthwhile to hand-sort the better ores for production purposes. However, samples sent to the General Engineering Company indicated that the highest concentration of zinc was about 30 percent, not much different from the concentrations that could be obtained by milling the unsorted ore.

Ore samples were then shipped to The New Jersey Zinc Company and the Blackwell Zinc Co. (American Metal Co. Ltd. subsidiary) in Oklahoma for evaluation using existing milling processes. American Metal tried processing in their own laboratory, and also sent samples to a company in Australia to try a process they were using. Results from all three tests proved unsatisfactory in removing much of the zinc concentrate from the ore, the best being about 45 percent by the Bradford process being used in Australia.

In the final analysis, Raymond concluded that the cost of milling the ore would be prohibitively high as compared with the price of zinc (7.5¢ per lb.) and lead (7¢ per lb.) at that time. He estimated that the maximum return per ton to be about $11.25 and the estimated cost to be about $9.00 for a gross profit of $2.25. The estimated 1 million tons of reserve ore would yield a profit of $2.25 million. Allowing for an additional 1/2 million tons of ore not already identified, the gross profit would be $3.375 million.

Goodrich-Lockhart was asking $2.4 million for the mine as it then existed. To this, Raymond added an estimated $1 million for construction of additional mining, milling, and transportation equipment for a total cost of $3.4 million. Measured against the estimated profit of $3.375 million, he had no choice but to recommend against his client purchasing the mine.

Goodrich-Lockhart was then faced with a decision as to what to do with the mine—one with a high profit potential only if a more efficient milling process could be found. No ore was shipped during this development stage, and the mine was allowed to fill with water.[4] On January 12, 1922, assets of the company, as they related to the Pecos River claim, were transferred from Goodrich-Lockhart to a new company incorporated in the State of Delaware under the name of the Pecos Corporation. David Goodrich was named President and a Mr. Hartwell, Secretary.

From 1922 to 1925, the Pecos Corporation bent its efforts toward finding a method of refining the complex ore. Solutions to the problem were sought in England, Europe, and Australia, as well as in the United States. In early 1925, Minerals Separation Company of San Francisco developed a process (known as differential flotation) that successfully floated lead concentrates away from the zinc. Although the process was not 100 percent effective, the search was over and efforts were then directed toward finding a suitable partner to help exploit the mine into a profitable venture.

4 The mine was extremely wet due to the constant inflow of water from Willow Creek. It had to be continuously pumped to keep it relatively dry. The pump used at that time was powered by a 160-hp wood-burning steam jenny that also produced the power for two hoists, one good for a depth of 600 feet.

chapter 2

✲ ✲

The New Owner

In December, 1923, a Mr. F. S. Norcross, Jr. presented to management of American Metal Co. Ltd. another study of the mine and its potential as a producing mine.[1] He was far more optimistic in his evaluation than Raymond had been. Rather than the $11.25 per ton value placed on the ore by Raymond, he placed it at $40 to $45 per ton. He reported that The Zinc and Lead Electrolytic plant located at Waldo, New Mexico, about three miles from the large anthracite coal mining operation at Madrid, would be an ideal place to have the ore smelted. He reported, "A happy combination of circumstances attends the situation, viz.:- the self interest of the Atchinson, Topeka and Santa Fe Railway; the advantageous railroad facilities; the coal and power situation; the future of by-product manufacture made possible by the suitability of the electrolytic plant processes and the availability of the anthracite coal (only large field in the Southwest); climate, labor markets, good roads, water and timber supply, all necessary to commercial success, are present."

Norcross later went on to say, "Metallurgical research and experimental work has been carried on for three years, and the process accepted herein was the result of detailed laboratory and pilot mill tests on representative shipments of the average ore." He does not, however, indicate who conducted the laboratory and pilot mill tests.

His study envisioned a mill site being established a short distance northwest of the town of Pecos; a narrow gauge railroad about 16-1/2 miles in length connecting the mine to the mill; a six-mile spur of the Santa Fe Railroad to the mill site; a coal-fired power plant located at either Rowe (about six miles south of Pecos) or at Waldo; and smelting of the milled ore at Waldo by the zinc/lead electrolytic plant located there.

It would appear that Norcross did not visit the site since he makes this statement, "The travel distance from Glorieta to the mine is approximately twenty miles over an *excellent macadam* road up the Pecos River Valley." Although the road to Pecos (about six miles) was the macadam surfaced highway US85, the road from Pecos to the mine site was graveled only to Valley Ranch (about two miles north of Pecos) and then a narrow, rutted wagon road for the remaining distance. At the time it was maintained by the U. S. Forest Service and often was not in condition to handle motor vehicle traffic.

---

[1] Norcross was a both a mining and mechanical engineer and an employee of Goodrich-Lockhart in their New York office. No doubt he was assigned by them to find a buyer for the mining venture. American Metal Co. Ltd. was cognizant of and privy to the Raymond study done three years earlier—it provided them with ample information regarding the mine's metallurgical potential. One can only presume, therefore, that Norcross was exerting every effort to dispose of the Pecos Corporation which was of little value as a producing mine at that time.

It seems probable that his report was based on inflated data in an attempt to generate interest in the mine. He closed his report by recommending that purchase of the mine be considered. Although no action was taken in 1923, when he again formally approached American Metal in early 1925 to enter into a partnership with The Pecos Corporation to exploit the mine based on the new milling technology, American Metal sent their own mining engineer to New Mexico to provide a more detailed report.[2] The following financial analysis and recommendation is excerpted from that report:

> We are offered a 52% participation in the property for $1,000,000. In addition to this we are expected to find funds to equip the property with mill, railroad (or aerial tram), power plant, and mine equipment for 600 tons per day. These funds will not be materially less than $2,500,000, and in addition to this sum, it will be necessary to spend up to $200,000 for pilot mill equipment and operation and exploration work underground. Furthermore, working capital will have to be provided.
>
> On the basis indicated, the 52% interest is not worth $1,000,000 unless the tonnage is at least 1,500,000 or the profits considerably greater than $5.00 per ton. The presence of 1,500,000 tons of ore is somewhat speculative as is also the probability of considerably greater profits. Unless we are willing to play these speculative phases, we are not justified in doing the business on the proposed basis.
>
> On the other hand, this is an attractive and very unusual property and through collateral advantages, such as the position of handling the tonnage of metal indicated, may be considered good enough to warrant the speculative features.
>
> In any event, the present proposed arrangement should not be entered into unless provision in the contract is made for the return of all money we advance up to the time of the start of commercial production, whether for operating or experimental equipment, experimental operations or exploration, before there is any dividend payment.
>
> If we could get a one year option on the property in which to prosecute exploratory work and metallurgical tests before the payment of the $1,000,000, it would be very well worth doing, and everything possible should be done to accomplish this result.
>
> If we could buy the 52% interest for $600,000, I would recommend it.
>
> Faithfully yours,
>
> (*Signature illegible*)

---

2 The report is dated July 25, 1925, and addressed to Dr. Otto Sussman, Vice President, American Metal Co., Ltd. Although the signature at the bottom of the page is illegible, other documents indicate H. S. Munroe was the author. This report implies that the Norcross report of 1923 was based on inflated data prepared by the Pecos Corporation. In it, he refers to and refutes several statements made in the 1923 report by stating, “The owners report....” then cites the statements made in Norcross’ 1923 report.

Although the exact details of the final transaction are not known, in October, 1925 American Metal decided to purchase a 51 percent controlling interest in the mine, and to move ahead as rapidly as possible to place it in operation. They expected to operate not more than seven years. However, it was decided to keep their intentions secret until all necessary land and water rights purchases had been consummated. Norcross was selected by David Goodrich and Heath Steele (President of American Metal Co. Ltd.) to serve as a liaison between The Pecos Corporation and American Metal, and to insure timely purchase of needed rights-of-way, land, and water rights. He was placed on the Pecos Corporation payroll to further insure secrecy of the venture. He moved to New Mexico and established an office and his residence at Valley Ranch, a western resort hotel.

While these negotiations were underway, The Pecos Corporation forged ahead with investigations of potential mill sites and transportation methods. An ideal mill site was identified in a small side canyon of the Pecos River (Alamitos Canyon) located four and one-half miles north of Glorieta and the Santa Fe Railroad. It provided ample space for the disposal of tailings from the mill with minimum danger of pollution of the Pecos River. The terrain over which a spur would have to be built was relatively flat.

Three methods of ore transportation from the mine to the mill were examined—trucking, narrow-gauge railroad, and aerial tramline. Trucking was discarded early-on because of the heavy tonnage involved, the poor, winding mountain roads, and the distance—twenty miles by road from mine to mill site. The narrow-gauge railroad was also discarded because of the extremely high initial cost and the probable objection of property owners along the Pecos River.

The remaining alternative, the aerial tramline, presented problems never before encountered in such a project. Cable manufacturers and tramway builders said it could be done, but never before had canyons been crossed with spans of over one mile in length. However, the idea was adopted, and survey crews were set to work laying out the best route. The work was done in the dead of winter with snow ranging from a few inches to three feet deep over most of the line. The ideal situation would have been a straight line between the two points, but that route proved impossible because the line would have had to cross a dude ranch and other unfriendly private lands. There was nothing to do but try again. On the fourth try, the route proved to be the most satisfactory from an engineering point of view, although not the least costly to install. The total length would be about 14 miles with an angle point located on the highest ridge. A double-control, double-drive station would be required at this point.

Francis Wilson of Wilson and Watson in Santa Fe, was Pecos Corporation's attorney of record and was retained to handle all legal matters pertaining to the transactions underway in New Mexico. At the direction of Norcross, Wilson began consummating purchases of numerous parcels of land required for the mine,[3] mill, railway spur, tramline right-of-way, water rights on the Pecos (needed to provide an ample water supply for the mill), etc. Because previous landholders had haphazardly transferred

[3] In addition to the mining claims, purchases also included the Chapin Ranch, the Simmons Ranch, the Marlowe Ranch, and the Cornell House. See Fig. 2 The Simmons property was purchased for a price of $13,000. One piece of property located in Dalton Canyon was purchased from Charles Dalton and wife, he being a son of Robert Dalton, leader of the infamous Dalton Brothers gang that robbed trains and banks in the late 1800's, and after whom Dalton Canyon was named. The elder Dalton had acquired the property while he was an Indian Territory U. S. Marshall before turning outlaw. The outlaw brothers used the property in the canyon for a hideout.

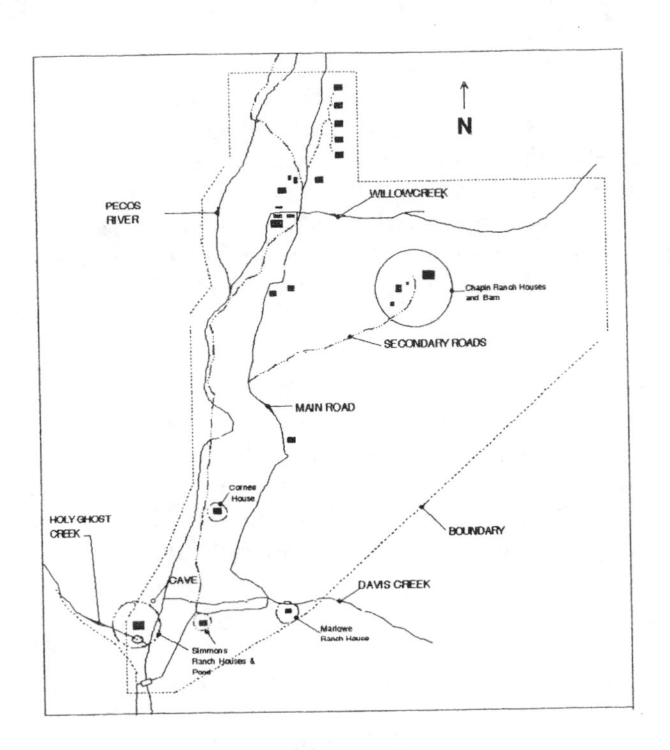

AMERICAN METAL COMPANY'S
PROPERTY BOUNDARIES & EXISTING STRUCTURES
December 2, 1925
FIG. 2

property from one to another, it was sometimes necessary to purchase small parcels several times over from all previous landholders to insure clear title to the property. Individual checks were made out to each pro-rata owner to insure no future litigation would occur. All new deeds thus obtained were made out to Mrs. Frederick (Beatrice) Pruyn who was Goodrich's sister-in-law by marriage to his wife's brother.[4] The multiple deeds were to be transferred to Goodrich on a single deed, and thence from him to the new company when and as deemed necessary. Norcross writes, "I had a local man purchase the properties in Mrs. Pruyn's name, she being a dummy." All this was done to keep secret American Metal's intentions. Even the Santa Fe Railroad was not initially advised of the true identity of their future client in the hope that they would agree to construct a spur at their expense, nor were they advised of the magnitude of the proposed operation.

On November 30, 1925, David Goodrich, as corporate president, filed papers with the County Clerk of Kings County, New York, dissolving Pecos Corporation and deeding its assets to The American Metal Company of New Mexico. On December 2, 1925, Francis Wilson, attorney for the corporation, filed incorporation documents for the American Metal Company of New Mexico with the Corporation Commission of the State of New Mexico. On Tuesday, January 19, 1926, incorporation papers of the American Metal Company of New Mexico, the deed transferring assets of the Pecos Corporation to the American Metal Company of New Mexico, and dissolution papers for the Pecos Corporation were all filed with the County Recorder in Las Vegas, New Mexico, County Seat of San Miguel County, in which the mine was located. The ***Las Vegas Optic*** reported the sales price as being $1,250,000. The American Metal Company of New Mexico was officially in business. Hereinafter, American Metal Co. Ltd. will be referred to as Amco Ltd. and American Metal Co. of New Mexico simply as Amco.

Following the filing of incorporation papers in December, J. B. Haffner, who had overseen the mine development work for Goodrich-Lockhart, was hired as General Manager of the new company. William Fisher from the Amco Ltd. office in Denver was moved to New Mexico as Chief Accountant. An office was set up in the old Pecos Corporation office at the mine-site; for lack of a post office at the site, however, the

---

4 In early 1936, Goodrich divorced his wife Ruth Pruyn Goodrich, and six months later married his sister-in-law, Beatrice Pruyn.

address of the company was established as The American Metal Company of New Mexico, Pecos Mines, Valley Ranch, New Mexico. Immediately, the two of them set about to hire key personnel to get the operation underway. Haffner hired E. C. Anderson as Chief Engineer, and Fisher hired Jimmie Russell as Chief Bookkeeper.

Anderson's first major job was to see that four antiquated log and plank bridges across the Pecos River were replaced. It was imperative that this work be finished as soon as possible; the transporting of heavy mine equipment and construction supplies to the mine could not be accomplished over the existing bridges. My father, by a stroke of good luck, would be the person hired to supervise that work.[5]

His first task was to hire a crew to replace the four bridges. While he sought and hired the crew, heavy timber-truss bridges were designed by Anderson, and materials were ordered from the Dockwiller Saw Mill located between Pecos and Cow Creek. Norcross had had the good wisdom to contractually tie up the mill's total production for all of 1926.

Designs were completed, a crew was hired, and construction began the first week of January on the bridge near Holy Ghost Canyon.[6] The work progressed, albeit slowly, during the freezing winter months. By mid-April, the bridges were complete; the roadway remained not much more than a wagon trail.

---

5 One evening early in December, 1925, Dad stopped by his barber to get a haircut. The barber told him he had read that day that Amco had announced plans to open a large mining operation on the upper Pecos River in North Central New Mexico. Eventually, they expected to employ over 800 people.

Dad, a carpenter, was working only part time and he correctly surmised that the mine would need carpenters from the beginning of operations. He and Mom discussed the possibility that evening; he left early the next morning to apply for work.

He traveled about 100 miles from our Mountainair home in central New Mexico to the mine site. He rode the train to Glorieta (via Albuquerque), then hitchhiked six miles to the little town of Pecos. He spent the night there and made friends with the owner of Harrison's Country Store. The next day, he walked the final 14 miles up the Pecos River canyon. There were no cars on the partly snow-covered, winding, mountainous road. He had been en route almost two full days, but the effort paid off. He was the third person hired by the new company.

6 This bridge still stands, but is now unused; it has been replaced by a parallel, modern, concrete structure. It was the first structure built by Amco, and is the only remaining as-built structure from the 1926-1939 era. At the Willow Creek Campground, there is a partial brick structure still standing. It was the mine's vault located on the south side of the mine office.

chapter 3

A Gilded Beginning

In early December, a decision was made to use the tramline[1] right-of-way for the high tension power line between the mill power plant and the mine. To protect it against forest fires or wind-blown timber, a right-of-way of 200 feet would be required. Long term leases were obtained from the Forest Service where the route would traverse public lands. Where private lands were to be traversed, the right-of-way was optioned and/or purchased, including rights-of-entry at convenient points from the road.

Anderson employed a survey crew to establish the exact route of the line, the profile of the right-of-way, and the control station angle point which was to be located on a ridge just south of Indian Creek, 21,468 feet southwest of the mine site. The distance to the mill site, almost due south, was 40,532 feet. The control point was 525 feet higher than the mine loading point, and 1,000 feet higher than the mill discharge point. Maps secured from the U. S. G. S. and the New Mexico Land Office proved to be so inaccurate that any thought of using them as guides had to be abandoned. From the angle point above Indian Creek, flags were sighted and set on the ridge tops south to the mill and north to the mine. A 300-foot wide strip was surveyed and contour-platted beginning at both the mine and the mill, and working toward the angle point; the flags atop the ridges were used as guides. Survey of the final route was completed in early February; clearing of the right-of-way began in early March.

When the route was fairly well determined in late December, representatives from various steel rope manufacturers and tramline builders were invited to study the proposed route, offer suggestions and make recommendations. American Steel and Wire Company sent their chief engineer; Leschen Company sent their assistant chief engineer; Roebling sent an engineer; and R. H. Riblet, president of The Riblet Tramway Construction Company of Spokane, Washington, went himself. All were shown the route, and were provided with all specifications and proposed details of the tram. They were asked to provide estimates and/or bids for furnishing only equipment and materials to be used by Amco to build the line, and also bids for furnishing a completed project.

For the next three months, the four companies provided many suggestions and proposals, including the use of timber towers, different types of cable, and various schemes of cable anchorages and tensioning. Amco came to the decision that they did not have the expertise to do the job, and would contract for a turn-key project by the

---

[1] In some quoted correspondence hereafter, the word *tramway* is used. I have chosen to use *tramline* throughout which is synonymous with tramway.

lowest bidder based upon specifications developed from all of the suggestions. Three bids were received in mid-March. Riblet's was the lowest at a cost of $313,000 plus freight costs estimated to be $15,000. Their bid included a guarantee that the tram would be completed and tested by January 1, 1927, in which event there would be a $3000 bonus and an additional $200 for each day completed sooner; if not completed on time, there would be a $3000 penalty plus an additional $200 penalty for each late day. A summary of the bid parameters follows:

Total length of line	62,000 feet.
Tram capacity	62-1/2 tons per hour.
Bucket capacity	1,250 pounds.
Tram speed	500 feet per minute.
Track cables -loaded side	1-3/16 in. dia. plow steel, smooth coil, 105 tons breaking strength.
Track cables -unloaded side	1 in. dia., plow steel, smooth coil, 58 tons breaking strength.
Traction cable mine side	3/4 in. dia., plow steel, 23 tons breaking strength.
Traction cable, mill side	3/4 in. dia., plow steel, 18.6 tons breaking strength.
Buckets	400, fitted with gravity type, self adjusting grip.
Towers	Bolted steel channels and angles with one coat heavy red paint.
Control point drive.	One, one hundred-hp motor for mine side drive and one, forty-hp motor for mill side drive.
Communication	Two, ten-gage galvanized iron transmission wires strung between mill and mine and supported by tram towers. Three field telephones and one each mine, mill, and control point phone.
Safety considerations	A control point solenoid-brake to bring the tram to a halt in the event of a power failure, or if the speed varied above or below 500 feet-per-minute by a predetermined amount.

The contract was awarded near the end of March, and Riblet immediately leased a fabrication site at Glorieta for the initial cutting, hole punching, and identification of tower and anchorage components. Engineers, foremen, and skilled mechanics were transferred from Riblet's Spokane facility to the site; all other workers were hired locally. Materials and equipment began arriving in Glorieta on a daily basis. Each structure's components were fabricated, and then moved by truck to an access point on the road nearest the structure's location on the line. From there, they were moved by mule pack-train to the specific location. Concrete footings were pre-cast at Glorieta and, along with the steel components, moved to the sites. Cement, sand, gravel, and water needed for on-site pour of the tension and control station's foundations also had to be moved in by mule pack-train.

Meanwhile, at the mill and the mine sites, work had begun to prepare both for full operation no later than December 1, 1926. At the mill, work was progressing at a better pace than it was at the mine. H. D. Bemis, who had been transferred to be mill superintendent from Climax Molybdenum (a Colorado mine in which Amco Ltd. had an interest) was pretty much given free rein by Haffner to get the mill into operation while he, Haffner, concentrated on the mine, the tramline, and a pumping plant on the Pecos River to supply water to the mill. Bemis had things moving along at a good pace.

By May 1, a five room building (eventually it would be the mill hospital) was completed and converted for use as a temporary office and staff headquarters. A tent colony for workers had been erected and a boarding house constructed and put to use. Grading for the mill building was about one-third completed, and excavation for the terminal ore bin and ball-mill floor were ready for trim-up before forming. Excavation for the power plant spray pond was complete and ready for forming. Grading and excavating had revealed that about half of the mill building would be located on clay; heavier foundations would be required than had been originally designed.

The Santa Fe Railroad people had moved along very rapidly with the spur line from Fox (a few miles east of Glorieta), and would probably complete their work in time for direct rail delivery of the milling and power plant equipment.

The situation at the mine site was a different matter—work was proceeding at a much slower pace. On the south side of Willow Creek, about half-way between the Katy Did and Evangeline ore bodies, the new four-compartment vertical shaft was being sunk. The elevation at the top of the shaft was 8,050 feet, 105 feet above Willow Creek, and 75 feet below the collar of the Evangeline shaft on the opposite side of the creek. The sinking had advanced to about 130 feet. De-watering of the old workings was underway using the steam-jenny for powering the old hoist, now being used for bailing, and to operate a 400-g.p.m. Cameron pump. A roadway to the new shaft was being constructed, and the foundation for a warehouse was being formed. Foundation excavations for a change house, an upper terminal ore bin, and a crushing plant were underway. However, by May 1, the schedule had called for the warehouse and the change house to be completed, and the new shaft should have been advanced to 200 feet.

Haffner had on March 15, signed a 90-day contract with Dr. Warren G. Smith to be the company doctor, both at the mill and at the mine. Sometime before the 90-day period expired, if either he or the company desired any changes, a new contract was to be drawn up for a period of two years. The essence of the terms of the initial contract were as follows:

1. The company agreed to provide an adequate hospital and a residence for the doctor and his family at the mine, as well as an office near the mine office. In addition, a smaller hospital was to be provided at the mill.

2. The doctor was to provide all services, including nurses, an assistant, furniture for the hospitals and the office, and all necessary medical equipment to operate the hospital. He was to pay for heat and electricity.

3. All employees were to receive free medical treatment for all illnesses, and for all injuries incurred on or off the job except that treatment for injuries received in automobile accidents when not on company business, or injuries received in fights or brawls, or resulting from drunkenness, and all treatment for venereal diseases were to be paid for by the employee at the same fees as would be charged for a family member of the employee.

4. The doctor was to provide office hours on a daily basis at the mine, and three times per week at the mill.

5. For his services to employees, doctor was to receive 50¢ from any employee remaining five days or less in the employ of the company, and $1.50 per month for each employee remaining more than five days. Both of these fees were to be deducted from the pay of each employee.

6. All family members of employees were to be provided medical care in the office at a rate of $1.00 per visit, and in the home during the day at $1.50 per visit and at night for $2.00 per visit. Night calls at the mill were to cost $5.00. All prescribed medicine was to be charged at no more than double the invoice price. Obstetrical cases were to be charged a flat rate of $25.00 to include delivery. In-hospital charges were to compare favorably with local charges (Santa Fe or Las Vegas) for the same or similar care.

On Monday afternoon, May 10, Joseph Matson, Sr. arrived at the mine office, having been sent there by Steele to report on construction progress. Steele had become concerned because he was not receiving timely reports from General Manager Haffner. Those that he did receive indicated limited progress, funds were seemingly being spent on items unrelated to the mining operation, and Haffner's estimate of the cost to get the mine and mill into operation had increased from $2 million to $2.5 million. Matson, at that time, was General Manager of The Climax Molybdenum Co.

Matson was appalled at what he found. Haffner, it seemed, was trying to build an empire for himself at Amco Ltd.'s expense, rather than working toward getting the mine into production. There was no effort being made to provide necessary additional housing for mine and construction employees. Nor was there much effort toward getting the new shaft sunk and necessary industrial buildings constructed. On the other hand, there was a great deal of effort being expended toward making the old Chapin ranch house a show place. Of the 16 carpenters then employed, six were working on the house and only ten were working on all other buildings. Four laborers were digging a five-foot-deep ditch between the mine and the house. The intent was to provide steam heating for the house from a boiler which would provide hot water for the mine change house. Only two laborers were employed digging the trenches for foundations for the change house. Haffner had had his office carpeted, and had purchased a handsome, solid walnut double-desk suitable for the highest paid corporate executive. Both the change house and the warehouse were being built at about double the size deemed necessary. It was also Haffner's intent to pipe steam to the office.

These, among other things, were Haffner's undoing. Matson called his superior at Amco Ltd.'s Denver office, Mr. H. L. Brown, to report what he had found. After conferring with Steele in New York, Brown called Matson to have him temporarily suspend Haffner, and take charge of the operation. The following Monday, Brown went to the mine to confront Haffner whom he then discharged. In Brown's report to Steele, he wrote:

> I didn't have a very pleasant time with Haffner, as you surmised, but those things, of course, are all in a day's work. In going through the correspondence, it is very evident that no one connected with the American Metal Company would have survived very long. Haffner had promised the mill superintendence to a man at Granby, telling him he could have the job when the mill was up and running. It was certainly a wise thing to have come to the decision you did, not that I want you to think that any of us who are now responsible have any exalted ideas of our own perfection, but we certainly aren't going to throw any more money away. We are going to make

some mistakes, and probably in your looking at them, you will think they were damn foolish, but I don't think we are going to make any that are awfully bad. Certainly any mistakes we do make aren't going to be gilded.

In other letters to other mine officials, he wrote:

> ....a rough estimate made by Matson and myself showed that we could figure on at least a reduction of $500,000 from Haffner's estimate. Also: Of course, there has been considerable money spent that to a certain extent was wasted, in that we could have accomplished the same results for far less money. However, even with this handicap, I feel quite confident that we are going to come within the estimate of two million dollars, and that we will have an efficient, economical operation.

Matson countermanded many of Haffner's orders, discharged an assistant Haffner had hired, a Mr. Rowbothan, who was occupying the other side of the double-desk, and arranged for the desk to be sold and a more modest one purchased. He was faced with many problems which Haffner had failed to resolve.

On May 18, in addition to the Haffner problems, he was faced with a sudden strike by the construction workers at the mill site. Labor laws being what they were in those days, over 100 men were discharged, and work came to a standstill. Matson reported it was the consensus of the foremen in charge that most of the strikers had been influenced by a few "agitators and undesirables" (some of whom had been personally hired by Haffner). He further stated that they would probably re-hire most of those fired as a new crew was being put together, and that care would be taken to weed out potential troublemakers.

The new shaft had been advancing at a rate of about one foot per day up to May 15, and was down to 134 feet after 139 days of work. The rate was increased to two and one-half feet per day during Matson's first week on the job, and was down to a depth of 155 feet on May 21. He set a goal of advancing to 400 feet by July 15, an average of about four feet per day.

After consultations with James Coulter, mine superintendent, and Cliff Hoag, chief mine foreman, a new plan of attack was developed to speed work in getting the mine ready for production. A temporary manway would be placed in the old Katy Did shaft below the 300-foot level in preparation for development and mining of the 400-foot level. A drift from the Evangeline shaft at the 400-foot level would be extended to a point directly below the new shaft, and a raise put up to the shaft above. This would speed completion of the new shaft. Also, as soon as the 400-foot level was de-watered, that level would be worked to prepare it for stoping; as soon as the new shaft reached that level, stoping could begin.

The sizes of the change house and warehouse were reviewed, and a decision made to cut the size of the change house by about 50 per cent. The warehouse building would remain the same size, since the foundation was already in place; however, an electric shop would be placed in the south end, and offices for the supply supervisor (Montgomery) and chief mine foreman (Hoag) in the north end. Also there would be a small change house for supervisors.

Poles for the power transmission line were ordered from Dockwiller so that they could be seasoned and ready for installation as soon as the right-of-way to the mill was cleared. A hard-wire telephone line to Pecos was being rushed to completion to cut out all cross talk being encountered on the Forest Service grounded-line presently in use.

As of May 20, employees at the mine totaled 192, at the mill, 70, and the staff at both sites, 38. The mill was short 100 employees due to the strike previously mentioned. The tempo of work had increased dramatically in just one week, but would be fraught with an ongoing problem of finding competent help for several months to come. Of the 300 employed on May 20, only 46 were married men, and 39 of them had families living elsewhere. There was a constant and serious problem of single employees getting a few weeks' pay, then quitting to go elsewhere where the weather was less severe. A few of the married men would become lonesome for their families, and also quit. Matson would solve the problem in ways never envisioned by the officers of the company when the decision was made to buy controlling interest in the mine.

chapter 4

✲ ✲

Matson Gets Things Moving

Under Matson's management, See Appendix F efforts were accelerated toward getting the mine and mill operational by January 1. Haffner's mismanagement had resulted in at least a two-month delay. The result was the elimination of any leeway in meeting the January 1 goal. The task seemed almost insurmountable; it required near-perfect timing and coordination if the target was to be met. As the weeks went by, numerous new problems kept surfacing which delayed one part of the project or another, thereby threatening to disrupt the overall schedule.

As previously stated, the mill at Alamitos had been moving ahead fairly well under Bemis' leadership, but completion was set back about a month by the strike of May 18. Work proceeded at a snail's pace until a full crew of 170 men was back in place.

A Mr. Humphrey had been hired to supervise the mill and power-house buildings construction. A Mr. Kennedy was hired to supervise installation of the mill equipment to be purchased from Minerals Separation Company, the firm that had developed the flotation process. However, the process had never been tried except in a pilot plant in the San Francisco laboratory. It had been planned that the mill would be operational by December 1, so that it could be operated for a few weeks to iron out the bugs before ore began moving on the tramline. That scheduled completion date had to be slipped to January 1.

By mid-July, work at the mill site was progressing in accordance with the new schedule. Mill grading was completed, and concrete flooring and foundation work were well underway. A limestone quarry above the mill provided the gravel for concrete used in the upper levels of the mill. Crews collected large boulders up and down Alamitos canyon with horse drawn wagons. Once collected, the boulders were hauled to a crusher at the lower end of the mill site to provide gravel for the lower levels.

The power-house spray pond was formed and ready for concrete; excavation for the power house itself was nearing completion, and ready for forming. There arose, however, a conflict between the Santa Fe Railroad crews that were working to complete the mill end of the spur, and the crews constructing the mill. The spur crews were blasting next to the power house, thereby partially filling excavations already completed by the power house crew. Conversely, the power house crew was blasting rock into the spur road bed already excavated. The result was an agreement to shut down power house work until the railroad blasting was complete—another short delay of about one week.

A six-room house with bath, costing about $2,500, was completed for Bemis and his family. Also nearing completion was a 16-room bunkhouse, and an eight-room bunkhouse just underway.

Originally, it had been planned to contain the mill tailings by constructing a retaining dike along the side of a gentle hillside south of the mill. A 45-foot-high trestle would be required to carry the mill tailings to the dike. Heavy rainfall during the summer months convinced management that there was a danger of flooding the tailings into the Pecos River resulting in law suits if any downstream environmental problems arose. The plan was changed in mid-August to dam Alamitos Canyon at a narrow point about one mile from the mill. The dam would be 50 feet high and would contain approximately 700,000 tons of tailings. This would handle the mill production for at least five years. The railroad spur was nearing completion, and the spur contractor was engaged to build the dam. Because all necessary equipment was on site, the cost would be about the same as that budgeted for the original plan.

Haffner had attempted, without success, to obtain a Pecos River pumping site from Valley Ranch to supply water to the mill for domestic, mill, and power plant requirements. He had eventually settled on a site about one and one-half miles farther away, over twice the distance from Valley Ranch to the mill. Matson would not accept that decision; he believed an arrangement could be worked out. He and attorney Wilson arranged to meet with Mrs. Miller, manager of Valley Ranch, to discuss the matter further. They worked out a deal for a site that would not interfere with the ranch operations and which would be of great benefit to both parties. In return for a one-acre site and a right-of-way for power and water lines over ranch property, Amco agreed to provide Valley Ranch with electricity at no charge for as long as the mine remained in operation. The one acre of land, according to the agreement, would be returned to Valley Ranch when, and if, Amco no longer needed the pumping plant.

This arrangement provided Valley Ranch with free electricity and with no restriction as to how much they could use. It would save Amco about $10,000 in initial material costs due to the shorter runs for the power and water supply piping, and there would be no out-of-pocket costs for the land and right-of-way. On June 22, the State Engineer granted Amco approval to remove up to three second-feet of water from the Pecos at Valley Ranch based on riparian water rights obtained by purchase of other lands between Pecos and the mine site.

Riblet's work on the tramline continued at a rapid tempo, but two problems arose which threatened to slow down the work. Riblet claimed there were errors in the survey which would have to be corrected before work could proceed on the tension stations at the mountain peaks; Amco found the quality of the sand and gravel being used in the tower pedestals to be inferior. After discussions with Riblet management, Amco agreed that there had indeed been some errors in the profile surveys and agreed to allow an additional 15 days before penalties would be imposed for failure to meet the completion deadline—the bonus for completion by January 1 remained unchanged. As to the quality of the sand and gravel, Riblet concurred it was not up to standard, and corrective action would be taken immediately to insure acceptable pedestals in the future. Those already installed would be replaced after the line was operational.

On Sawyer Creek, (about two miles south of the angle point of the tram) Riblet had put up another small fabrication shop to take care of on-site changes found to be necessary in tower and tension station components. By July 31, they had submitted bills for about one-half ($148,000) of the contract cost. All towers between the mine and Macho Creek were erected, and eight of the ten tension stations were complete. The double control station was about 20 percent complete. A two-mile stretch between Sawyer Canyon and the control station was in actual operation hauling construction materials to the station.

Approximately five and one-half miles north of Pecos, in the middle of some high cliffs on the west side of the river, there was an outcropping of Early Pennsylvania bituminous coal.[1] Goodrich-Lockhart had secured a patent on the coal vein, and had opened a mine to provide fuel for operation of the steam-jenny at the Evangeline shaft. Known as the Pecos River Coal Mine, the operation had been shut down when mine development ceased in 1922.

Amco reopened this mine as soon as it became necessary to fire up the steam-jenny to provide power for de-watering the mine. By the end of June, production of coal had reached about 100 tons per month. The cost of the mining operation (mined by contract) was quite expensive compared to the cost of coal on the open market. The lower transportation cost for the Pecos River coal, however, balanced out the total cost, and a decision was made to continue with the operation until the tram was operating. At that time, the plan was to purchase open market coal, move it by rail to the mill, and transport it to the mine on returning tram buckets.

The goal of having the new shaft sunk to the 400 level by mid-July was not achieved. The problem was the difficulty in finding skilled miners, and in keeping them on the payroll when found. Most skilled miners were married men, and simply did not want to stay without their families. Three solutions to the problem were initiated:

1. A contractual arrangement was established whereby a skilled miner could put together a crew and contract to accomplish a given amount of work by a specified time at a specified rate per foot of raise, shaft sinking, or drifting. The crew members were guaranteed a rate of pay at not less than that paid to non-contractual employees. If successful in the contract, the miner and his crew could earn a good bonus for their extra effort.

   Amco continued to pay the miner and crew their regular wages, and the total sum paid was deducted from the contractual earnings at the end of the specified period. The balance was paid to the miner who in turn divided it with

---

[1] On the top of the ridge above this outcropping, and about one-half mile north where Dalton Canyon joins the Pecos, there was a wooden cross implanted by a religious sect known as the Penitentes. At Easter each year, a new cross was emplaced by the sect in a ritual to atone for their sins of the previous year. They would climb to the point in a single file, each stripped to the waist. The last person in the file carried the cross strapped to his back, and a whip in one of his hands. With this he lashed the back of the man immediately in front of him. That person would do the same to the person in front of him and so on up the line. They arrived at the top with their backs bloody and welted.

The sect owned an acre of land at the point where the tramline's south side tension station for the Dalton Canyon over-crossing had to be located. Norcross had had to obtain 38 signatures to secure clear title to the acre and, after 1925, the ritual moved to some other location. The last cross remained in place for several years.

his crew according to their agreement; Amco was not privy to the details of those agreements. Contractual rates were $5.00 per foot for sinking and raising, $7.00 per foot for drifting. The wage scale was $4.00 per day for miners and $3.00 per day for muckers until July 1, when they were raised to $4.50 and $3.50 respectively, all based on an eight-hour shift. Contracts were for labor only with the company paying all other costs.

2. A miner, who owned and used his own car, was contracted to recruit other miners from Arizona, southern New Mexico, and if necessary, Mexico. He was paid a specified sum for each recruited miner who stayed on the job for a fixed period of time. Every two weeks or so, he returned to the mine with three to six men. Some stayed, others returned to their families elsewhere after a few weeks of work.

3. A decision was made to begin construction of houses for married employees, and to cease building bunkhouses for single employees. Five additional eight-man bunkhouses were under construction on an open site above the mine. The engineering staff developed plans and site-plats for construction of 36 two-room and 14 three-room houses. Work began almost immediately on construction of ten of the two-room "cottages," as they were called, just below the road to Cowles. Each was a 12- by 24-foot, wood-frame structure covered with diagonal wood sheathing and mineral-surfaced tarpaper. Interior walls and ceilings were faced with a one-quarter-inch "beaver board" with joints hidden under one-quarter by two-inch lath. Interiors were painted with *Kalsomine,* a water based zinc-oxide mixture containing glue and a colorant.

The road between Valley Ranch and the mine was in the process of being widened and improved with a base course of gravel. Many places were too narrow to accommodate two-way traffic and, although traffic was light, accidents had occurred far too frequently. Although the road belonged to the Forest Service, Amco absorbed all of the cost.

Upon completion of de-watering in late July, as much work as possible was devoted to preparing the Katy Did ore body for mining. Several ore chutes, raises, drifts, cross-cuts, and main and waste haulage drifts on both the 300 and 400 levels were being worked concurrently. The main haulage drift that had been developed by Goodrich-Lockhart had to be timbered due to irregularities and swells in the foot-wall that had occurred during the years the mine was under water.

The power line along the tramline was completed in early October, but transmission of power to the mine had to await completion of the power plant. Meanwhile, some power was being provided by two diesel-fueled generators at the mine, purchased and installed for use as a stand-by power source in the event of an emergency after the mine was in full operation. The steam-jenny was shut down on August 15 and with it, also, the Pecos River Coal Mine. The plan to transport coal on returning tram buckets would not need to be implemented.

The crushing plant was completed in late October and erection of buildings over the plant was begun. Test runs of the plant were started using diesel-generated power. Tests indicated that additional 54-inch rolls and a ten-inch bulldog gyratory crusher would be required, and these were placed on order. Although they could not be in place by the scheduled January 1 production start-up, the initial minimum production could be processed in the plant as it had originally been designed.

Equipment and material for the Pecos River pumping plant were placed on order in August after title to the site was signed by Valley Ranch. Work began on clearing the right-of-way for the water line and power transmission line. A three-room house was begun for the pump plant operator and family.

Although the total output of the Dockwiller Saw Mill had been contracted for by Norcross, many delays were encountered due to Dockwiller's limited capacity to produce. An average of over one-quarter-million board feet of lumber per month was required, but Dockwiller continued to place some on back-order. Additional lumber had to be purchased from Santa Fe Builders Supply Company and Santa Fe Mill and Lumber Company. The skip and cage guides in the shaft were made from clear, kiln-dried Sitka cypress which had to be special ordered from Weyerhaeuser in Washington state. The cost was over three times that of the ponderosa pine lumber obtained from Dockwiller, or about $60 per thousand board feet.

Because Willow Creek flowed across the principal shear zone of the Evangeline ore body, a decision was made to dam the creek about one-half mile east of the shear, and to divert Willow Creek into a flume to bypass the zone. To insure total entrapment of the water, the dam had to be emplaced on bed-rock; the job was far more difficult than anticipated in that the bottom of the canyon was filled with sand and gravel to a thickness of about 16 feet. By comparing the amount of water having to be pumped prior to fluming with the amount pumped after fluming, it was determined that there had been a total inflow of Willow Creek water of over 100 gallons per minute. However, the inflow of water from other sources was still considerable; continuous pumping would still be necessary, but at a decreased rate.

chapter 5

✲ ✲

January 1, 1927—The Venture Begins

Month after month, the most challenging problem during the entire construction and development program was that of maintaining an adequate and skilled labor force both at the mill and the mine. Most of the skilled miners, carpenters, millwrights, and mechanics were married men and few would stay for long without their families. The premise had been from the beginning that single men housed in bunkhouses and fed in mess halls would make up the majority of the work force, but few were found with the necessary skills who would stay for more than a few weeks. The turnover rate was 30 to 40 percent each month.

The goal was to have no less than 80 men working above-ground and 290 men underground by January 1. On December 13, Matson reported that the the above-ground force was 80 as planned, but that the underground force was only 154 and declining, since more men were leaving than were being hired. He warned New York management that the pace of start-up operations would be considerably less than expectations.

Of the work force that was in place, the best workers, and the ones most likely to stay on the job, were those miners, both Hispanic and Anglo, who had been provided with houses for their families. Out of 40 who had been favored with houses, only three had quit during the previous two months. Of the remaining work force, over 100 had quit for a turnover rate of almost 40 percent per month.

Despite the employee problems, all phases of the construction program began coming together about December 1, although mine development work had not progressed as fast as had been hoped, not only because of the shortage of skilled miners, but also because the compressors for supplying mine air were not fully operational. They could be operated only when diesel-generated power was available; power from the mill was yet two weeks away.

The headframe and the hoisting system were in place, but also operating only at limited capacity due to the lack of power. The same was true of the crushing plant; it had been tested and found to be ready to begin crushing when power was turned on.

Riblet had slowed down somewhat on installation of the tram due to the loss of two good foremen and because of bad weather. However, they continued to insist the tram would be operational by January 1.

All mill equipment was installed and ready with the exception of some minor items that were not important for initial operations. Two essential items were missing, however—power and ore from the mine. Power transmission lines to the mine and the pumping plant were in place and the power plant was ready to operate, but cooling water from the pumping plant was needed before the generators could function. The pumping plant itself was ready, but two failures had occurred in the pipeline when the pumps were first tested, and these were in the process of being repaired.

At 10:05 AM on December 14, one of the generators began producing power using cooling water produced by temporary wells. Pumps at Valley Ranch were started a few minutes later, and the storage tank and spray pond at the mill began to fill. By mid-afternoon, the generator was operating at full capacity, and switches were thrown to send power to the mine. By late afternoon, all equipment at the mine was up and running, and the diesel generators were shut down. It would be nine days before power was delivered to the mine and mill communities.

At the mine, a large Christmas tree, perhaps 25 feet tall, had been set up in the school yard and strung with Christmas lights, not the colored kind of later years, but plain white incandescent bulbs. The company had purchased a Christmas gift and a Christmas stocking filled with an orange, an apple, an assortment of nuts, and hard Christmas candies for every child in the eighth grade or below. On Thursday evening, December 23, switches were thrown to send power into the community and the tree lights began to shine. After Santa Claus distributed the gifts, everybody returned home to turn on their own electric lights—the need for gasoline or kerosene lamps had ended. Earlier in the day power was supplied to Valley Ranch and to residents at the mill.

On December 28, the tramline began operating at diminished capacity. Placed on line that date were 60 buckets, and ten more would be added every day until all 400 were on line. Outbound buckets were only partially filled to insure a safe beginning operation. A number of minor operating problems did surface. There had to be numerous adjustments in tensioning of track and traction cables. In one instance, traction cables were adjusted too tautly, with the result that several empty buckets were lifted off the track cable. One tower was slightly damaged, but quickly repaired.

On December 31, Matson concluded that Riblet had fulfilled the contract to be operational by January 1 and authorized payment of the $3,000 bonus. The $200 bonus for each day of operation in December was not paid. Plans were made to transfer ownership of the tram to Amco on January 15, along with an operating crew that had worked on installation and testing of the system. A Mr. Dahlgren, a Riblet employee, would become the tramline foreman in charge.

The crushing plant operated on waste rock from the time power was available on December 14 through December 31. In those 15 working days, 1,500 tons of waste rock had been hoisted and crushed, and subsequently trammed to the mill for its trial runs. The hoist operated without problems and the crusher with only a few minor adjustments necessary.

On December 31, the power plant, operating with one of the 1,500 kW generators on line, produced 7,400 kWh; about 11 tons of coal were consumed. At maximum load based on full operation of mine, mill, and pumping plant, it was estimated over 20,000 kWh would be required each day, consuming around 30 tons of coal. The cost was figured to be 1¢ per kWh.

In the mine itself, 20 stopes and 20 pillars in the Evangeline ore body were being developed and prepared for mining; 20 stopes were already developed and partially prepared for mining in the Katy Did ore body. To produce 18,000 tons of ore each month, it was planned to mine 20 stopes, to fill 20 stopes, and to develop 20 stopes concurrently.

On January 1, 1927, the mill began operating on waste rock to provide training for all operating crew members, to provide 1,200 tons of sand to fill thickener bottoms, and to provide a fine silt to fill and seal cracks in the wooden tanks used in the milling process. Matson sent the following telegram to his boss in Denver:

> WIRING STEELE GOODRICH TODAY THAT MILL IS TURNING OVER ON WASTE FROM MINE DELIVERED BY TRAMWAY DURING TRIAL RUN OF LAST FEW DAYS STOP TRAM IS NOT OPERATING CONTINUOUSLY DUE TO MINER (sic) LINE TROUBLES WHICH COULD NATURALLY BE EXPECTED STOP HOPE TO SEE YOU NEXT WEEK AND OUR BEST WISHES FOR A HAPPY NEW YEAR STOP J. T. MATSON

On January 3, Steele replied to Matson by letter as follows:

> I was happy to receive your telegram announcing the starting of operations at Pecos. The fact that you have been able to make a record in construction gives me a great deal more pleasure than winning the bet which I made with Mr. Goodrich that you would make it.
>
> You will naturally have some worries in smoothing out your operation, but I confidently expect you to make just as good a record in your operations after you get started.
>
> We all wish you and the Pecos organization a very happy and prosperous new year.
>
> Yours very truly,
>
> Heath Steele

Mr. C. E. Lewis from Metals Separation Company in San Francisco arrived at Alamitos on January 19 to check on the operation of the new plant and the equipment they had supplied. He reported back to his superiors that the mill was functioning just as they had predicted from their laboratory test and that he found the mill had been constructed precisely as they had recommended. The following is his brief description of the milling process:

> The mill proper has two sections, each with a capacity of 300 tons. The ore crushed to approximately two inches is fed from the fine ore bins by pan conveyors to one of the three Marcy 75 ball mills in closed circuit with a Dorr classifier. The overflow going to two 12 by 12 conditioning tanks equipt (sic) with Devereaux agitators. The discharge from these tanks going to the 5th cell of a 16 cell, 18 inch Sub A Machine for the lead float. The lead machine tailings go to an 8 by 12 tank for lime conditioning and from there to a second 16 cell Sub A Machine for the zinc float. This machine produces a finished zinc concentrate and a final tailing. Both these products go to tables for recovering of some of the gold iron values and also serve as pilot for the flotation operations. Intermediate thickening in the zinc circuit is installed in the first section but not in the second altho provision has been made for this installation.

Reagents. The reagents in use now are essentially the same as used in our large scale tests namely Cyanide, Cresylic acid and Xanthate in the lead circuit, and lime, Copper Sulphate, pine oil and Xanthate in the zinc circuit. These reagents are fed by automatic feeders with mechanical distributors on the reagents where distribution is required. Results to date indicate that little regulation of reagents is required in spite of fluctuations in metal content of the ore.

In late February, Steele and David Goodrich visited the operation and hosted a dinner at the La Fonda in Santa Fe to celebrate the start-up of mining and milling operations. Bemis was one of those who attended, and in writing to Steele about the event, he apologized for having to have dental surgery at that time and being able only to take liquid refreshment through a straw. He referred to it as the "sowbelly dinner." Perhaps it was barbecued pork spareribs.

In Matson's progress report on March 9, concerning February's first full month of operation, he referred to certain inaccuracies in two figures on recovery rates of zinc. One report showed an 87 percent recovery, while another showed it to be 94.6 percent. He ascribed the difference to inaccuracy in assays run by inexperienced samplers, men who were new to the trade and with, as yet, insufficient training. All this indicated that it was unknown how much of the mined zinc was sent to the tailing pond—13 percent or 5.4 percent.

In any event, the report shows that out of 10,305 tons of ore milled during February, 3,211,302 pounds of zinc and 368,598 pounds of lead were sold to the smelters. In addition, there were 371 ounces of gold, 233 ounces of silver, and 26,707 pounds of copper. The prices at that time were $20.67 per ounce for gold, $.5828 per ounce for silver, $.0727 per pound for lead, $.1264 per pound for copper, and $.0665 for zinc.

He further reported that development work in the mine was still about two months behind expectations due to the early labor shortage. Since all work after January 1 was being charged to production and not development costs, the additional development thus charged was seriously affecting the cost of production, and a resultant decrease in profits. He predicted the development would be up to schedule within a few weeks. At that time, the true cost of mining could be determined and the profits shown on reports would be more accurate.

The first serious problem on the tramline occurred on February 11. On line were 385 buckets, one-half carrying full loads of ore—the heaviest load yet placed on the system. Tension station #5 began to give way, and it appeared that a long shut-down of the system would become necessary. Dahlgren, who had had considerable experience while employed by Riblet, suggested some timber reinforcements be installed to forestall a total failure of the tension station. With the repairs in place, operations resumed on February 14, and 523 tons of ore were transported that day. The average transported up to that time had been 328 tons per day.

Riblet was immediately notified of the problem, and they dispatched an engineer and construction foreman to investigate the cause. After a thorough inspection of the station, their conclusion was that all tension stations were under-designed and would need to be modified. However, they concluded that, if the tram could be operated only at night and their crews could work during the day on the modifications, no continuous shut down would be necessary. They agreed to make the changes as rapidly as possible.

The mill had been operating with both units, but the decrease in ore transportation resulted in a reduction back to one unit while repairs were being made. During that period, modifications to the roll crushers at the mine would be made, resulting in a finer crushed product and an increased output of the ball-mills in the milling process. A decision was also made to install seven additional shaker tables for tailings to insure maximum recovery of gold.

With these changes completed, Matson predicted that full production of 600 tons per day would be reached within two or three months, and maximum recovery of metals would be achieved.

chapter 6

* *

Mining Practices

For the technically inclined[1], it seems here appropriate to provide a detailed account of the physical characteristics of the mine and the methods used to exploit it. [2]

Physically, the ore bodies consisted of irregular, disconnected, often overlapping lenses of sulphides in the shear zones. These lenses would pinch out abruptly, either horizontally or vertically.See Figs. 3 & 4 Mineralization often replaced all of the schist, in which event the ore body would have a diorite foot and hanging walls. In other places, only certain bands of the schist were replaced and both walls would consist of schist.

Both the ore bodies and the walls were found to be quite loose and required heavy timbering. The ground was found to run freely in many places, resulting in the possibility of serious caving. However, if initial movement could be prevented, support was not too difficult. Unfortunately the presence of so much water made it more difficult to support the ground due to its high talc content. For this reason, it was always important to develop levels below stoping areas in which the water of the stoped area could drain and be pumped to the surface.

Prospecting for new and continuing ore bodies was accomplished by drifting on the principal ore shoots along their strike. If the ore body pinched out, an attempt was made to follow the most promising of several sheared zones. The method was a gamble at best, since ore occurrences were erratic. For this reason, ore bodies once located would be explored by driving ahead the sub-level (which constituted the sill floor of the stopes) considerably in advance of the main level drift. By working in this manner, the main level drift could be placed to much better advantage with respect to the turns, widenings, or pinching of the ore.

---

1 Numerous mining terms are used in the following two chapters which can be found defined in the Glossary of Mining Terms beginning on page 178.

2 Certain portions of the next two chapters have been paraphrased in whole or in part from *The U. S. Bureau of Mines Information Circular 6368* authored by J. T. Matson, Sr. and C. Hoag and published in 1930.

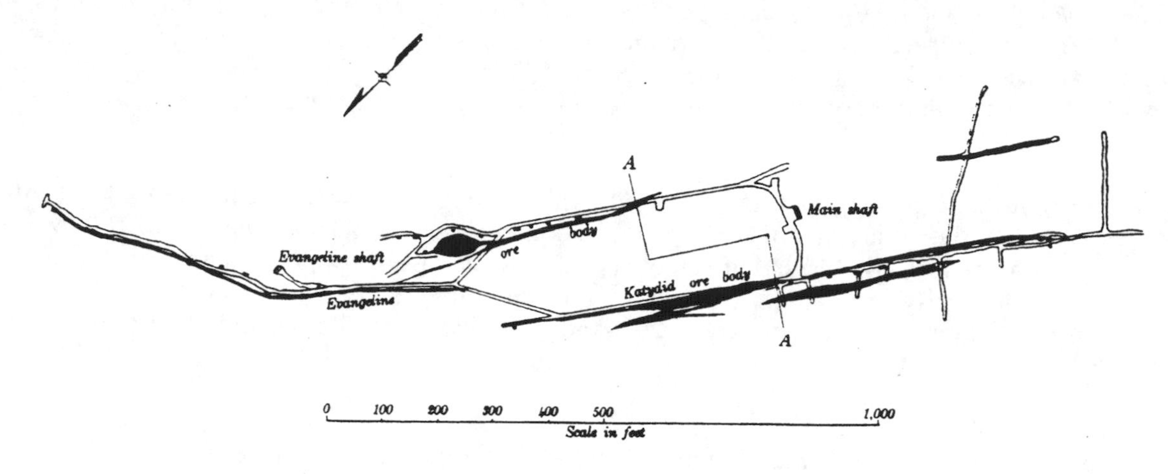

Figure 3—Plan of 400 Level

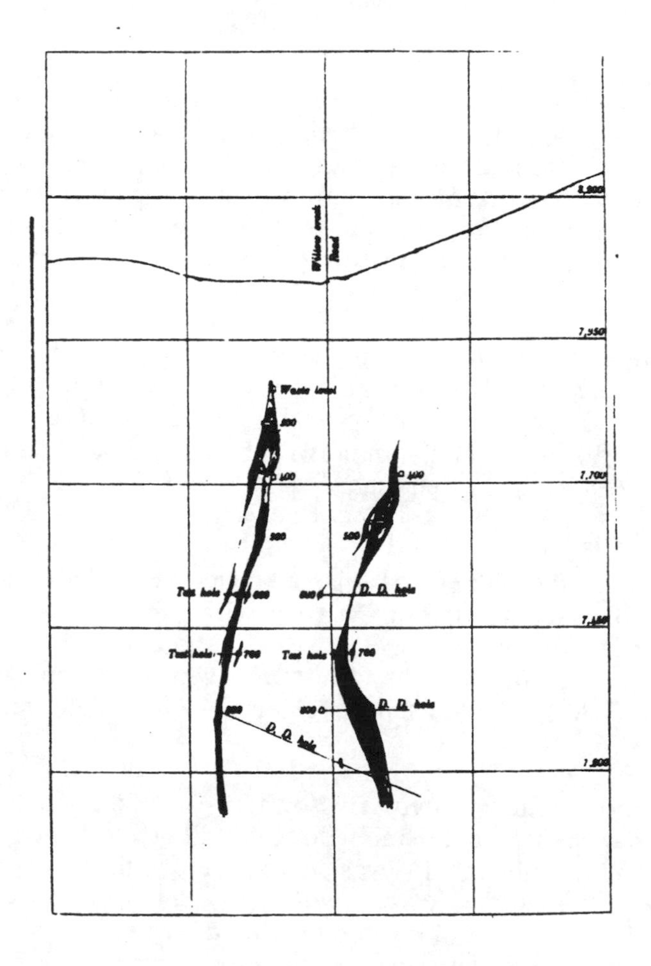

Figure 4—Vertical Section Through A—A, Fig. 3

Shear zones were crosscut or diamond drilled at frequent intervals. Walls and downward extensions were explored by drilling from underground, whereas extensions along the strike were explored by drilling from the surface. Walls were also explored by deep-hole drilling with Leyner machines and sectional steel, the holes ranging between 50 and 100 feet in length.

All underground openings that showed mineralization were sampled by cutting channels. Usually, this was done with a hammer and moil. However, if the ore was hard, a compressed-air hitch-cutter was probably used. The channels were cut normal to the schistocity wherever possible, and the length of the channels included in individual samples made to correspond with bands of high, medium, and low-grade ore or bands of waste.

The grade of the ore was usually easily determined by inspection. As a result, sampling was unnecessary for stoping purposes. Sampling, however, was usually done along the boundary lines of each floor of a stope previous to filling. The tonnage produced in each stope was determined by survey and checked by the car tally which was first corrected to agree with the tons milled.

A core recovery of about 75 percent was obtained by diamond drilling. The sludge was always assayed, but experience showed that the sludge assays were quite inaccurate. Therefore, when the sludge assays showed mineralization, the cores were split, one-half being assayed, the other half kept for permanent record. This procedure precluded overlooking low-grade mineralization in the drill cores. All cores were stored in a basement area located beneath the mine office building and were readily available for inspection.

Ore reserves were computed from closely spaced vertical transverse sections. Grade estimates were based on the average of all assays made on a given block. No account was ever taken of possible or probable ore. Therefore the ore outline was extended only a short distance into unknown ground, and was a relatively accurate measure of known reserves at any given time. See Appendix B

As previously indicated, the mine was opened with two shafts and an adit level. The main, a vertical four-compartment shaft, with levels at 100-foot intervals, was midway between the two ore bodies. All ore and waste hoisting was done through this shaft, as well as part of the handling of workers and supplies. The old two-compartment Evangeline shaft was used exclusively for the handling of workers and supplies. See Figs. 5 & 6

The main shaft consisted of two five-by-five-foot skip compartments, one four-by-five-foot sinking compartment, and one three-by-five-foot manway, ladder, and pipe compartment. See Fig. 7a. When the shaft was being sunk, the two skip compartments were bulkheaded about 20 feet below the lowest working level, and broken rock was allowed to accumulate on the bulkhead. The sinking compartment was lagged off from both the skip and manway compartments from the surface to the bulkhead.

Horizontal development work consisted of crosscuts from the main shaft to the veins of ore, and drifts along the vein, generally in the footwall. Raising consisted of chutes and manways from a level to the sill floor of the stopes, 20 feet above, and fill raises driven from level to level upon starting the initial stope section in a new ore body.

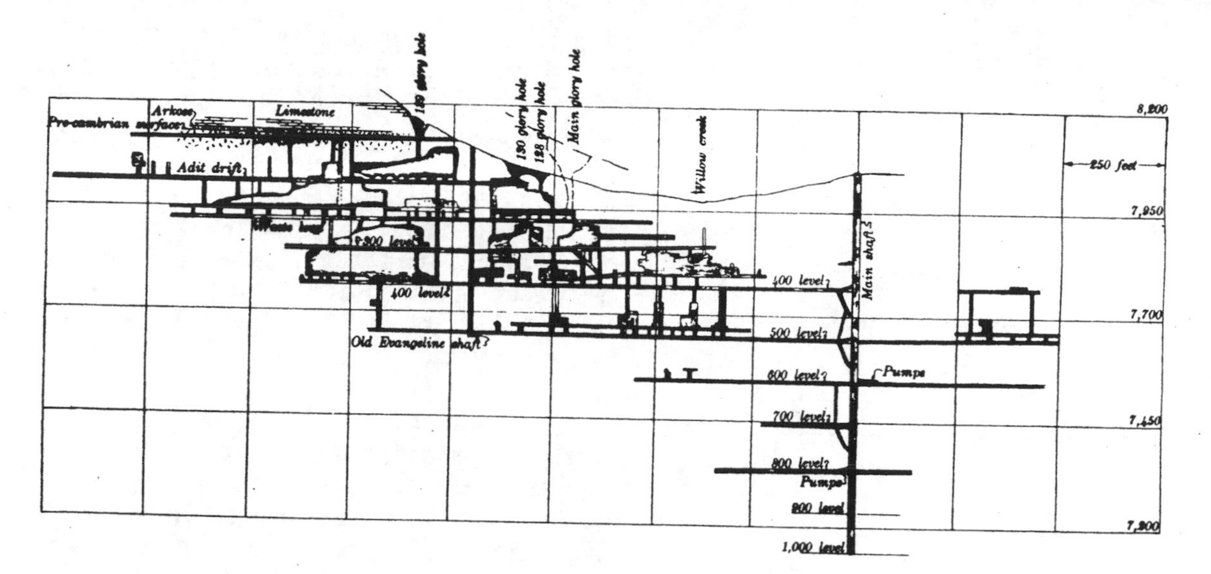

Figure 5—Longitudinal Projection Along Evangeline Vein (Yr. 1930)

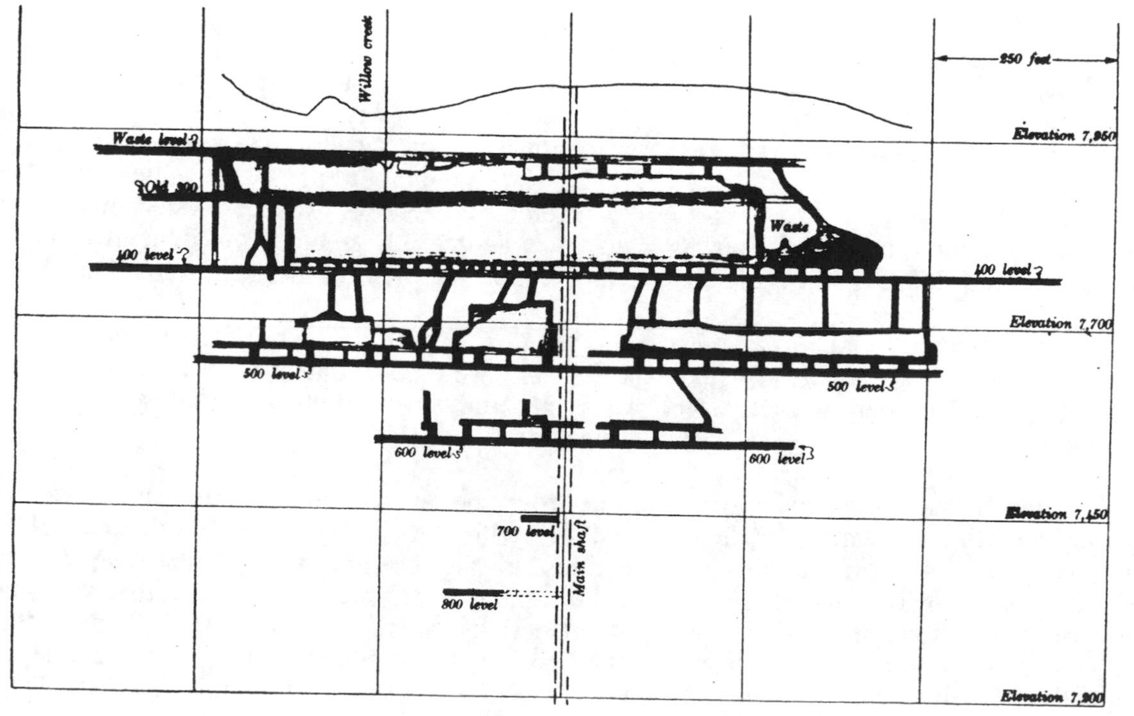

Figure 6—Longitudinal Projection Along Katy Did Vein (Yr. 1929)

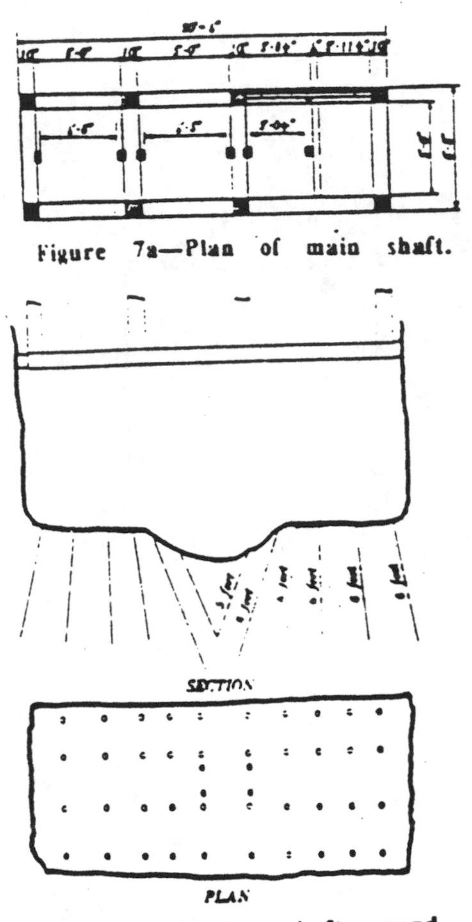

Figure 7a—Plan of main shaft.

Figure 7b— Typical shaft round.

There was no particular effort devoted to obtain speed in the development work once mining was underway, since the rate of development was dependent upon the ease of waste and/or ore disposal. Development work above the sections being stoped was carried on whenever waste was needed for stope filling. Lower level development depended upon the amount of ore or waste that could be hoisted in the main shaft in addition to the handling of men, stoped ore, and supplies.

A one-half-yard sinking bucket, with safety crosshead, was installed in this shaft. Two buckets were used, one being raised and dumped while the other was being filled. Broken rock from the shaft sinking was hoisted to a pocket on the 200-level. From there it was drawn off for stope filling. The bucket was dumped by fastening a chain hung from a wall plate into a ring on the bottom of the bucket. Upon being lowered, the bucket would tip and empty the broken rock onto a trap door. This door was hinged to a wall plate so as to swing into the shaft at an angle of 40 degrees. The dumped ore would slide on this trap door into the 200-level pocket.

The blasting round used for sinking was of the center V-cut type. See Fig 7b The usual round consisted of 44 holes. Experience had shown that a large number of holes gave the best results, since the rock was broken smaller and there was less damage to the timber due to the decreased amount of dynamite in each hole. Because of the loose nature of the ground, timbering had to be kept close to the bottom.

An 85-pound jack hammer was used for sinking. The heavy machine had been adopted because, at times, ribs of hard rock were encountered making necessary the use of the bigger hammer. Drill steel was one and one-eighth-inch hollow round, lugged for a Leyner chuck.

Rounds were blasted with two to four boxes of 40 percent strength, special low-freezing gelatin dynamite. Firing was done with ten delays of #8 electric delay blasting caps, using a 440 volt circuit. Connections to the bottom were made with several hundred feet of blasting wire carried on a reel. For safety, the shaft boss kept the key and the fuse plugs of the detonator in his pocket. Connections at the bottom were made with two lead wires suspended on stakes clear of the ground. All black wires were connected to one lead, all red wires to the other, but no connections were made until all men were ready to leave the shaft.

During shoveling, drilling, and timbering, light was provided by several large light bulbs attached to a reflector. However, before loading of dynamite began, the light transmission line was detached and hoisted to prevent the possibility of stray currents reaching the blasting harness.

Usually, a six-foot depth round would require eight hours to drill, load, clear the area, and blast; 24 hours to shovel; and another six hours to timber, install guides and ladders, and to lower the pump back into place in preparation for the next round. Shoveling was difficult because the broken rock tended to cement itself, and picking was required for each shovel-full.

Coeur d'Alene lagging was used because it strengthened the sets when blasting close to the timber, and it had the advantage of being easily replaced. About 50 gallons per minute of water had to be pumped from the shaft bottom in addition to the amount hoisted during mucking. Concrete water rings were placed about every 100 feet in the shaft. They served as timber bearers as well as to catch water coming down the shaft walls and timber. Water caught by the ring was raised about 20 to 25 feet to the next level by the use of eductors, a device using water under pressure as a motive force.

Air operated displacement pumps were used at the bottom of the shaft, the exhaust air being discharged into the water column. This avoided freezing of the exhaust valves

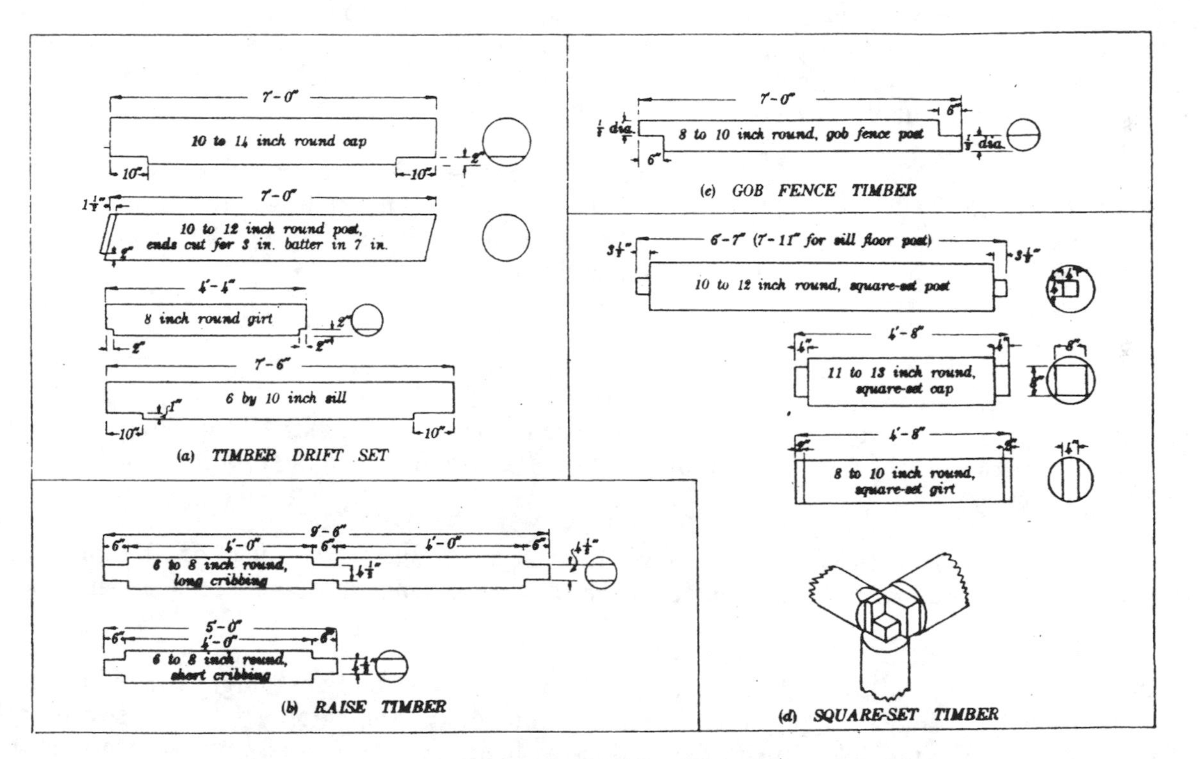

Figure 8—Detail of Mining Timbers

and reduced the noise level. Two pumps were used and interchanged after each round of blasting. The one not in use was hoisted to the next level above and thoroughly inspected for possible repairs. When deemed to be serviceable for another round, it was lowered back to the bottom to be ready for the next round. An electric centrifugal pump was always held in reserve should an unusual amount of water be encountered.

Crosscuts from the shaft to the veins were driven seven-by-eight feet in cross section. Haulage drifts were timbered seven feet high by five feet, four inches at the top and five feet, ten inches at the bottom in the clear. See Fig 8a. Haulage drifts were usually driven in the ore following the footwall. A toe-cut blasting round was usually used in drift work. From twelve to 16 holes were drilled for each round, but in the loose ground, which usually prevailed, the back holes were not blasted. If found necessary to make room for timbering, they would be blasted later. The practice avoided many serious runs or cave-ins.

All shoveling was done by hand. No regular daily cycle of operations was possible due to the very irregular nature of the ground. Therefore, the crews on both day and night shifts had to perform all of the work—drilling, timbering, or shoveling, as found necessary on that crew's shift.

From the main haulage drifts, cribbed manways and chutes were driven to the stope sills about 20 feet above the level, at intervals of 25 to 60 feet, depending upon the width of the vein. Raises were cribbed with round timber, about six inches in diameter See Fig. 8b, forming two compartments, each four by four feet in the clear. Short pony sets were placed over the drift sets where a raise was to be started. The pony set caps were long enough to extend to solid ground on one side of the drift. These caps formed the sills on which raised cribbing was set.

Fill raises from the stope sills to the level above were generally stulled, about four by six feet in section. Three-inch planks were nailed to one side of the center stulls to form a manway. In some places the raises had to be cribbed, in which event they were larger, having two four-by-four-foot compartments.

Three types of drills were used in the mine. Leyner type drifters mounted on three and one-half-inch bars were used in drift work and sometimes in stope sills. Heavy jackhammers, as previously stated, were used for shaft sinking and lighter ones in the stopes for breaking ground and blockholing. Hand-rotated stopers were used for raise work and in stopes.

A repair shop for drills and air hose was located near the collar of the main shaft. A "drill doctor" visited all working places each day to attend to minor repairs and to keep track of the machines.

Three types of hollow drill steel were used: one and one-eighth-inch round for Leyners and shaft sinkers, one-inch quarter octagonal for stopers, and seven-eighth-inch hexagonal for light jackhammers. A single 14-degree taper cross bit was used in all cases. Starter bits were two and one-quarter-inches in diameter. Length changes were 18 inches for Leyner and jackhammer steel and 12 inches for stoper steel.

A daily average of 360 pieces of steel were sharpened each day by one drill sharpener and two helpers. They also tempered the steel.

A 3,000 cubic foot per minute compressor near the shaft collar supplied air at 100 pounds pressure, which was sufficient to maintain a working pressure of 85 to 95 pounds at the drill. Drilling water from the domestic water supply was piped to all locations.

Underground, two types of dynamite were used—40 percent strength special low-freezing gelatin and 25 percent strength straight gelatin. The main magazine was located at the surface and only enough explosive for daily needs was kept underground. A nipper at each level was in charge of the level magazine and distributed the prepared fuses and dynamite to the working places.

Number eight electric-delay detonators were used in all wet places, in high raises, and in long straight headings. In relatively dry places, number six tetryl caps were used with a triple-taped waterproof fuse. About half of the tetryl caps were treated with a cap sealing compound. No stemming was used, and each miner did his own blasting.

chapter 7

* *

More Mining Practices

Several stoping methods were used in the mine for maximum ore recovery and to maintain safe working conditions. The method selected for any particular body or vein was dependent upon the characteristic features of the particular ore vein being mined at any given time. Stoping methods were sometime changed because of newly encountered conditions. The ore shoots in general were irregular and not continuous, pinching out and giving way to others. On occasion, two to four parallel veins were separated by only ten to twenty feet of barren, loose schist, or blocky diorite. These parallel veins were sometimes mined simultaneously, leaving the waste in place; but most often the waste had to be mined along with the ore, hand-sorted, and gobbed back into the stope.

Since most of the veins contained many low-grade or barren schist bands and diorite intrusions, the mining method adopted had to provide for waste sorting to avoid serious dilution of the ore. Approximately 20 percent of the stoped ground had to be hand-sorted and rejected in the stope as waste.

The method selected had to be such that runs from the walls could be checked quickly and, that walls could be prospected at close intervals to discover parallel ore shoots. The faulting which occurred during or after the ore deposition had left gouge in the walls up to several feet thick which had to be guarded carefully to prevent runs.

The surface had to be kept intact to minimize water intrusion into the mine. The Pecos River, with a normal flow of 200 second-feet, paralleled the outcrop of the veins at a distance of only 400 feet, and the dip of the veins was toward the river. As has been previously stated, Willow Creek, with a normal flow of 20 second-feet, crossed the shear zone at right angles at its widest point, and had to be flumed across the zone.

Cut-and-fill and square-set-and-fill methods of mining were used about equally. The choice between them depended upon the character of the ore and walls where stoping was started. Strong walls and/or ore favored the cut-and-fill method, however, the method sometimes had to be changed after stoping was carried up one or more floors, due to changing physical conditions of the walls or ore.

Pillar mining by underhand stoping was used on occasion because it proved to be slightly less costly than the square-set method. However, it did require considerably

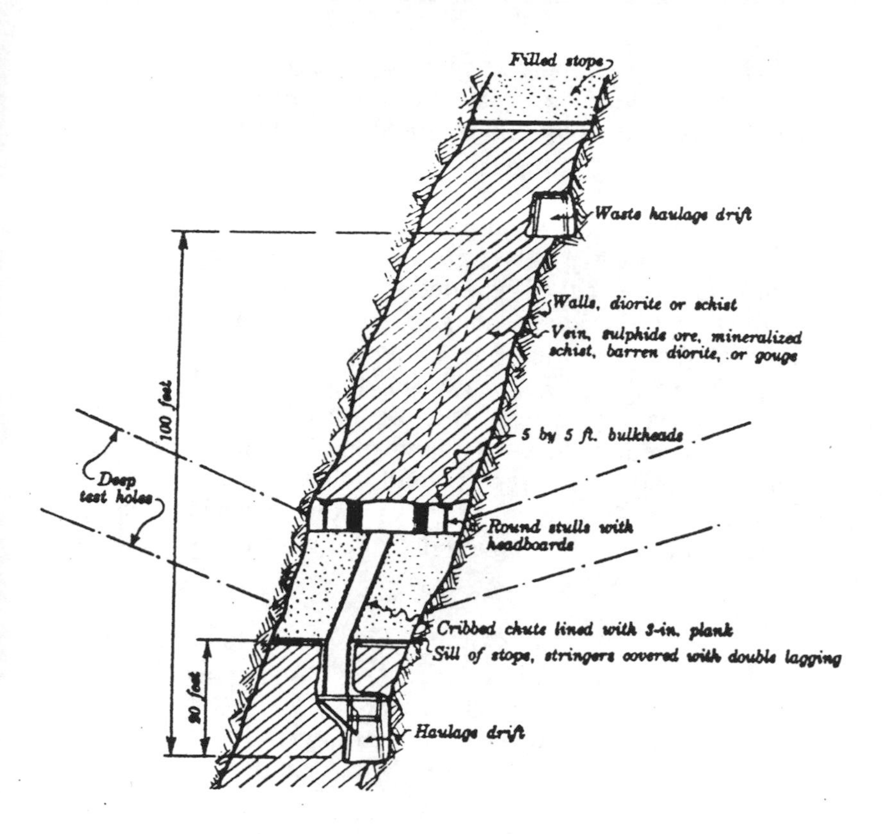

Section Across Vein Through Cut-and-Fill Stope

closer supervision since large unfilled areas tended to result in caving if not very carefully worked. Caved stopes proved to be more costly to reopen than a filled stope.

Shrinkage stoping proved to be possible in only one stope near the top of the ore body. The ore and walls were stronger there than lower in the ore body, and caving would have had no serious effect on future operations should it have occurred.

Inclined cut-and-fill was tried and rejected as a mining method, due to the tendency of the ore and walls to cave and run, with consequent flooding of the stope and danger to the workmen below. Also, the method made it difficult to erect stulls, cribs, and bulkheads on the inclined floor. These disadvantages outweighed the saving effected by gravity handling of ore and wastes on the inclined floors.

Horizontal cut-and-fill was the method of choice whenever possible. From the main haulage drifts, two-compartment raises were driven 12 feet above the back of the drift, and the stope was silled out the full width of the ore. See Fig 9 Ground conditions at that point determined the method to be used, whether cut-and-fill or square-set. Even though many stulls and bulkheads might be required due to loose ore and walls, and as much timber would be required as in square-setting, lesser skilled labor could be used for cut-and-fill, resulting in lower costs overall.

The ore bodies were mined out in sections, the length of the sections dependent upon the width of the vein and on ground conditions. Initially, in very wide veins, sections were from 15 to 60 feet in length with 15-foot pillars between. Later practice was to cut initial stopes 30 foot long, spaced at 100 to 150 feet. Extraction chutes were located at both ends of the sections. After the initial stopes reached the level above, new sections were started on one or both sides, using the former extraction chutes for fill raises. Pillars were located only where two stope blocks came together.

The stopes were mined upward by taking successive seven-foot cuts from the hanging to the footwall, following each cut by filling. Stope end lines were timbered off from the solid by means of vertical gob posts See Fig 8c spaced five feet apart and laced with lagging. The posts were tied back into the fill by means of old cable.

In narrow veins, stopes were silled several hundred feet in length with chutes spaced 30 feet apart. Every other chute was provided with a manway. Fill raises were driven in ore 30 or 60 feet apart, dependent upon the width and grade of the ore.

Stopers were used for drilling in all stopes where possible. However, in many places, the ground was so loose that it was more satisfactory to drill flat holes with lightweight jackhammers. Because of the nature of the loose ground, it was necessary to provide prompt support for the back, in which event breaking ground resolved itself into making room for timbers, rather than to producing tonnage. Little heavy blasting was ever done, and blasting near the end of a shift was held to a minimum.
Before blasting, a shoveling floor of two-inch planking was laid on the fill. The plank was used again on the next floor if not broken, in which event it was used for lacing the outside of cribbed raises. All shoveling was accomplished by hand due, to the necessity of hand sorting. Cribbed extraction chutes were lined inside with three-inch planking. This lining was always repaired after each cut in the stope, and sometimes during a cut to prevent runs of waste fill. Tops of the chutes were covered with grizzlies of round timber spaced about ten inches apart.

Back and floor pillars were left standing until all stoping was completed on the level below. They were then removed by square-setting in short sections, usually by placing two sets along the vein and extending them from wall to wall. The general rule was to retreat toward natural outlets. As a rule, drifts were kept open permanently. When they were no longer needed for future work, they were filled tightly when the stopes were filled.

The same general scheme of operations was used when square-setting, except that manways, extraction raises, and gob lines were provided by lacing off the square-sets. Square-sets were made of round timber, were five feet square and six feet, eight inches high, except that sill floor sets were eight feet high. See Fig 8d The caps were stronger than the ties, and were therefore placed normal to the strike which was the direction of most pressure.

Waste for filling stopes was obtained from development work on the levels, from prospect work driven into walls from the stopes themselves, from sorted waste, and from glory holes on the surface. A raise from the upper level, known as the waste level, was connected to a glory hole at the surface. A bulldozing chamber with a 12-inch grizzly was located in the raise about 40 feet above the level. However, during the first few years of operation, prospect and development workings furnished sufficient waste for all requirements, and the glory holes were not used.

The movement of ore and waste underground was accomplished with 40-cubic foot rocker-dump ore cars, and 18-cubic foot side-dump waste cars, both riding on 16 pound rails at 24-inch gage. The waste car See Fig 10 was small enough to fit in the lift cage, so as to be carried to the surface and manually moved on rails to the dump site, or to be carried to one of the waste haulage levels for movement to stopes for filling. The reason for using the side-dump cars, rather than end-dumps, was to avoid the possibility that, over time, some of the walls of waste haulage levels could have squeezed together enough to minimize clearance. Whereas the side-dumps could be emptied in place, end-dumps, on the other hand, had to be uncoupled and turned 90 degrees to be dumped, an operation requiring considerably more clearance.

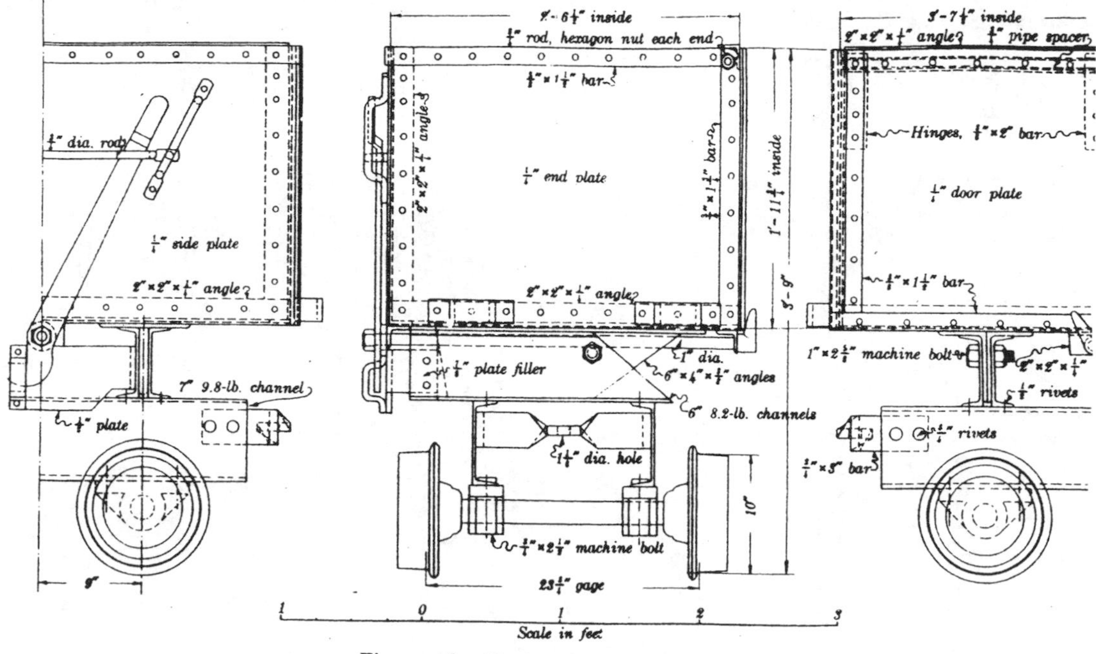

Figure 10—18 Cubic-Foot Side-dump Car

Storage-battery locomotives were used to move the cars on the older levels when the distance from the shaft was such that mechanical power was justified. For the shorter distances in new level development, the tramming was accomplished by hand. Dual batteries were used in the locomotives when large tonnage was moved.

Ore pockets were cut into the main shaft at odd-numbered levels. Even-numbered levels were served by ore by-pass raises from the ore pocket to the next level above. See Fig 11 Ore pockets had a capacity of 300 tons each. Steel grizzlies made from 90-pound rail spaced at ten inches in the clear were installed over both the ore pockets and the ore passes.

The two skip compartments of the main shaft were equipped with combination skips and cages, with the cages permanently attached beneath the skips. See Fig 12 Skip capacity was three and one-half tons each, with the two operating in balance; that is, one was being lifted while the other was being lowered. The hoisting cables operated off twin cable drums connected in-line, one having been wound clockwise and the other counter-clockwise. This arrangement placed a lift burden on the hoist no greater than the weight of the ore in the skip and/or men and materials in the cage. From the 700 level ore pocket, about 100 tons per hour could be hoisted. Hoisting speed was 720 feet per minute for both men and ore. The main hoist had a rope pull of 20,000 pounds.

The main shaft sinking compartment was equipped with a single drum electric hoist with a rope pull of 12,000 pounds. When shaft sinking was not underway, the hoist operated a small skip used to handle waste from all development work. This waste could be dumped into a waste pocket on the 200 level or into a surface waste bin to be subsequently loaded into side-loading waste cars and manually trammed to the dump. The hoist at the old Evangeline shaft had a rope pull of 7,000 pounds, operating a single-deck personnel and materials cage.

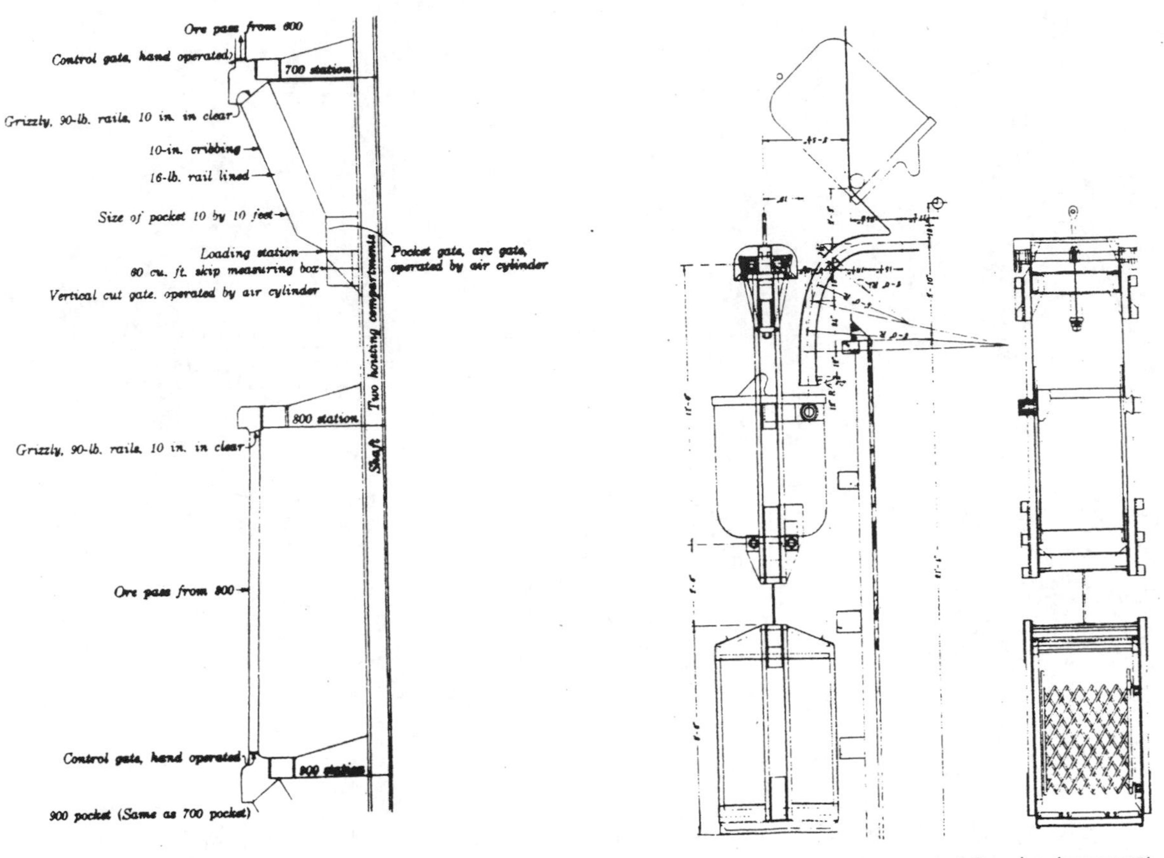

Figure 11—Ore Pass and Shaft Pocket Arrangement

Figure 12—Cage, Skip, And Dumping Arrangement

The ore bin at the shaft collar had a capacity of 100 tons. From this bin, the ore was fed into an 18-inch by 30-inch jaw crusher, the output of which fell onto a conveyor belt. Some waste was removed from the conveyor belt as it moved the primarily crushed ore to a 10-inch gyratory crusher. From this it flowed by gravity into 20- by 54-inch rolls. The resulting 1-inch product was conveyed up to the top of a 1,500-ton storage bin at the tramway loading station, and subsequently moved by the aerial tram to the Alamitos mill. Operation of the tramline is covered in detail in the next chapter.

As before stated, the mine was extremely wet and large quantities of water had to be pumped to the surface. In the main, centrifugal pumps were used to lift the water, but one 450 gpm vertical-plunger displacement pump was used at the 500 level. Centrifugal pumps were of 500 to 1,000 gpm capacity driven above the 800 level with 440-volt, direct-connected electric motors, and below that level with 2,300-volt, direct-connected electric motors.

As the mine increased in size, the volume of water increased almost proportionately, and additional pumps had to be added. Diamond core-drilling in exploratory work opened new channels for inflow of water, and it was a constant battle to keep these plugged. In one instance water pressure at one hole measured more than

160 pounds psi at the 1,000 level. Between 1927 and 1932, the amount of water pumped increased by more than ten times, from about 250 gpm to over 2,500 gpm.

Initially, the water was brought to the surface through two eight-inch-pipe water columns located in the manway/ladder/pipe compartment of the main shaft. In 1933, it became necessary to churn drill a hole from the surface to the 900 level, its location being about 700 feet south of the main shaft, and directly over a smaller #2 shaft which connected the 900 and 1,200 levels. A 12-inch-pipe water column was installed in this hole and further extended to the 1,200 level down the #2 shaft. Two 900 gpm pumps were installed to lift the water directly to the surface from the 1,200 level. These pumps operated continuously for the next five and one-half years at 1,800 gallons per minute. In that span of time, they produced over 5.2-billion gallons of water, this in addition to the water being produced through the eight-inch columns which did not operate continuously.

Because of the numerous openings into the mine (two shafts, an adit, and three glory holes), plus numerous ore and waste passes and manways connecting levels one to the other, ventilation of the mine was no problem and mechanical ventilation was not required with but one exception. When a new level was opened, small fans moved air into the ends of crosscuts and drifts through canvas tubing until such time as connections were made to the level above. The interior of the mine was always quite cool, and during the winter months could get down to freezing. Air doors had to be installed in many crosscuts and drifts to prevent cold air currents from freezing the water and air supply lines.

chapter 8

✲ ✲

Tramline Operation

The tramline-operating crew consisted of ten men on each shift—three each at the mill unloading terminal and the control station, and four at the mine loading terminal. In addition, there were two oiler/inspectors who worked only on the day shift. The procedure at the mine terminal was for one man to inspect each empty bucket after it was automatically released from the traction cable, oil the trucks, and push it around to a loader. The loader filled the bucket at one of the four crushed-ore storage bin chutes, then pushed it to the third man who was the bucket launcher. Every 36 seconds a light would flash and a gong would sound to signal the launcher to send another bucket on its way. This was necessary to keep all 400 buckets (traveling at 500 feet per minute) on-line, and to insure even loading of the system.

The launcher, obviously, had the most difficult and grueling job; each 400-pound bucket was loaded with 1,250 pounds of ore, more or less, and there was no time for rest. To spread the workload, and to allow for break and lunch times, the crew of four men rotated the three tasks so that each man worked seven and one-half hours, had two 15-minute breaks and a one-hour lunch period during the nine-hour work shifts.

The rail on which the buckets rode inside the terminal was ingeniously designed to make bucket movement, as well as bucket launching, less difficult. The first ten feet of the rail was inclined eight inches upward from the incoming track cable. For the remainder of its length on this incoming side, the rail was level, but the "U" turn around to the outgoing side was declined two inches. Between the "U" turn and the launch point, it was again level, and from that point to its intersection with the outgoing track cable, it declined another six inches.

The speed of the incoming buckets was such that, when they were automatically released from the traction cable, their momentum carried them up the incline onto the level portion of the rail, where they slowly came to a stop. The oiler would perform his task, then push a bucket onto the decline portion at the "U" turn. Gravity carried the buckets around the turn to the loading chutes. The loaded bucket was then pushed on the level rail to the launcher, who would in turn push it, on cue, onto the final decline. As the bucket rolled down this decline, gravity increased its velocity to the point that its speed was synchronous with the traction cable at the instant when the two met and the automatic grip was activated.

At the control station and the unloading terminal, the crews worked in much the same manner except that no oiling or inspection was required; a three-man crew could handle the job. At the control station, there were two launchers and one system operator. The three also rotated tasks to provide the heavy side launcher relief from that grueling and monotonous job. At the mill end, there was an unloader who released the bucket from an upright position to allow the ore to dump into storage bins. The next person swung the bucket back into its upright position and locked it in place, and the third person launched the empty bucket back on its way to the mine. As with the other two stations, the men rotated their tasks.

The bucket suspension system was a patented design by Riblet. Suspended from in-line, two-wheeled cable-riding trucks, the buckets had replaceable track wheels. The suspension arms pivoted on the truck assembly to permit the truck to assume the angle of inclination or declination of the track cable. A lever arm with a cast iron 32-pound truncated cone weight at the outer end operated the cable gripping system. It was mounted between the two wheels and when rotated upward, caused the cable grip to release; when rotated downward, it activated the grip. The iron weight served two purposes: 1) it rolled along a horizontal spiral cam thereby raising or lowering the arm, and 2) its weight provided the force to securely grip the cable. Its 32-pound weight was multiplied to a force of over 3,000 pounds per square inch at the cable grip.

Spacing of the tram towers and tension stations was dictated by the terrain. Near the crest of a mountain, spacing was often as little as eight feet, whereas the distance at the Macho Canyon over-crossing (between two mountain tops) was slightly over one mile; at Dalton Canyon, it was just under one mile. Height of the towers and the distance between buckets and the ground below was also a function of the terrain. Occasionally, the distance was as little as three feet, whereas at the Dalton Canyon over-crossing, the distance to the creek below was just under one-half mile.

To insure minimum downtime due to cable failures or other malfunctions of the system, a crew of two men, classed as oilers/inspectors, rode the buckets every day to tighten bolts, to oil moving parts, and to inspect for cable wear or other potential problems. They were equipped with portable telephones with which they could call the control station or one of the terminals whenever the need arose. One of the two began the two daily rides at the mine and the other began at the mill. When they met at mid-point, they would reverse direction and return to their respective starting points. If they detected a flaw in the system, and it was deemed an emergency, they could dismount at the next tower, connect the portable telephone to a telephone pair strung at the tops of the towers, and call the control station operator to shut down the system. Otherwise, the problem would be entered in a log book, and a repair crew would be dispatched to correct the problem the next day.

These men became very proficient at mounting and dismounting the buckets at a tower or at tension stations. It was essential they have no fear of height nor be subject to vertigo. However, the skills were not gained without incident. In mid-April, one of the men attempted to catch a ride from tower No. 4 back to the mine. He missed the bucket, but caught the traction cable and hung on for about 900 feet before dropping some 50 feet to the ground. He broke his leg and sustained major internal injuries, bruises, and abrasions, but managed to crawl back to the tower to catch another bucket back to the mine. He was hauled out of the bucket and rushed to the hospital. Doctor Smith gave him little chance of recovering. However, in a couple of weeks, he began to rally and eventually fully recovered, but would not ride the buckets again.

Another of these oiler/inspectors, Rupert "Bud" McInteer, also had a harrowing experience on the tram. One evening, he failed to show up for supper at our house where he was a boarder. No one had seen him or knew where he might be.

As it turned out, around mid-afternoon, the mill-end oiler/inspector had called in to have the system stopped due to high winds. The operator failed to advise his relief at the change of shift that Bud was somewhere on the system. It happened that he was left swinging in the wind over Macho Canyon. Over about three-fourths of the line, he would have been in a bucket low enough to drop to the ground and call in his position. The drop where he was stranded was over 1,000 feet.

Around dusk, the wind died down, and Bud decided to walk the traction cable to the nearest tower by grasping, hand-over-hand, the track cable above. However, he soon faced a problem—the traction cable sagged out of reach of the track cable. He solved this by using his tool and regular belts, buckled together, to hold the two cables within reach. He inched the belts along ahead of himself, drawing the two cables together, as he slowly made his way to the next bucket. Fortunately, a full moon that night provided enough light for him to see what he was doing. There were two buckets between him and the nearest tower when he started, and he took a well-earned rest in each.

He was resting in the second when his whereabouts was ascertained, and the system was started up to allow him to get to a tower. Around three a.m. he called in to report his location. He had to make his way down the mountainside to Macho Church, where someone had been sent to pick him up. He arrived at our house just before breakfast, and recounted his harrowing experience. After breakfast, he returned to his room in the bunk-house and slept until breakfast the next day. He was not affected by the experience, and went back to work the next day. Another traumatic experience, of which I was a part, did affect him quite seriously about three years later. This experience is related in Chapter 21.

The second serious mishap on the tramline occurred on Sunday, May 8, 1927. A bucket left the track cable on the light side (northbound) at tower No. 69 a short distance south of Dalton Point, caught the tower and uprooted it. The traction rope broke on that side, resulting in runaways of buckets both north and south of the point. The broken ends were separated by about one mile, and forty buckets were damaged as they plowed into each other—some damaged beyond repair. The sudden release of tension resulted in cable whiplash at both the control station and the mill. Three men received serious injuries at the control station, while one at the mill received minor injuries. One of the three did not survive. This accident placed a severe burden on the six-bed hospital, since four beds were already occupied. Rooms for six additional beds were added the following year

A power line pole was struck by a falling bucket and destroyed; power was interrupted at Tererro for over twenty-four hours, although the stand-by diesel generators carried the industrial load, and work in the mine continued. Ore bins at the mill were depleted in about forty-eight hours, and the mill had to be shut down.

Repair of the cable had to be made at the Dalton Canyon span. New cable was ordered from Leschen Company in St. Louis on Monday, and was received in Glorieta on Thursday afternoon. Repair materials had to be delivered to the site on rented pack horses—the only ones available were underfed and weak. A pack train of five, pulling cable up the steep mountainside, was dragged off the Dalton trail by the cable; one was killed and the others severely injured. Tramline downtime lasted eight days and mill downtime, seven. Stretching and expansion of the traction cable during the next three weeks resulted in two additional downtime periods to provide time for length readjustments.

This mishap underscored the need for a highly skilled repair crew and special techniques for making repairs. The problem apparently was caused by high winds blowing an empty bucket into the tower as it passed by, so a system was devised to guide buckets around towers even if blown toward them by high wind. However, it was always the practice to shut down operations if heavy wind came up along the system and was detected in time.

A Caterpillar 30 tractor was eventually purchased to eliminate the need for the horse pack trains, and other special cable splicing equipment was obtained. The repair crew of six men, when not repairing damage along the line, repaired or constructed new buckets, constructed access trails and roads to every section of the line, and otherwise busied themselves with repair and maintenance tasks. When there was a major shutdown of the system, the repair crew was supplemented with men from the operating crews.

As the weather warmed, tramline speed began to surge at the mill end; the warmer the weather, the more pronounced the surging. Mr. Riblet personally visited the site to try to determine the cause. The end result was to change from a rigid tensioning system to one which could be adjusted to take up the slack as the cable expanded. In addition, Anderson designed a system whereby the traction cable could be replaced with little downtime at minimum of expense.

Track cables had to be changed about every other year, and in some cases even more often. Special temporary clamps were installed when up to three cable strands wore through and broke; the entire cable had to be replaced if more breaks occurred. During the first two years, the track cable was turned 120 degrees every three months according to accepted theory and practice. However, this was discontinued when it was determined that the cable lasted longer if left alone as originally installed.

The only other serious injury ever sustained on the system was not by a workman, but by an eleven-year-old boy. Between the dairy and the mine, there were two places where the buckets rode about three feet above ground. Anyone could jump onto a bucket while running along beside it, stand on the handle used to empty the bucket, and hold onto its vertical supports.

Many boys found it to be great sport to catch rides this way, and jump off when the distance to the ground began to increase. The boy, Norman Littrell, failed to jump in time one day, and when he did jump, he broke his arm. The next day word went forth from management that the father of any boy caught riding the tram would be summarily dismissed. Another form of boyish recreation came to an abrupt end.

chapter 9

✲ ✲

Bargaining for a Rail Spur

The required railroad spur to the mill was completed by the end of August, 1926. From the time the mill site was selected in August, 1925, until February, 1928, when a final contract was signed, the spur had been a matter of much negotiation, scheming, paperwork errors, and second-guessing between Amco and the Atchison, Topeka and Santa Fe Railroad. The first meeting between the two companies occurred on September 10, 1925, in Chicago. Present were Norcross from The Pecos Corporation; Steele and a Mr. Bien (a specialist on railroad transportation) from Amco Ltd.; and W. B. Storey, President, and F. B. Houghton, Traffic Manager from AT&SF.

Steele had obtained a letter of introduction to Mr. Storey from E. N. Brown, Chairman of the Board of Directors of the St. Louis-San Francisco Railroad Company. The letter did not identify Steele as President of Amco, rather simply as a mining company executive interested in discussing the possibility of a spur to a potential mining venture near Glorieta, New Mexico. Storey and Houghton were very cautious and conservative in their reaction to Steele's proposals. They mentioned a cost of $40,000 per mile for construction. Bien had previously estimated the cost at $25,000 per mile, based on his experience with other spur constructions.

Storey was reluctant to make a commitment until provided a bond insuring that their expenditures would be covered should the venture not materialize as proposed. He indicated AT&SF had lost money on prior mining ventures which had folded before AT&SF had recovered their costs. Steele suggested that Amco install the spur and that AT&SF reimburse Amco at a rate of $4.00 per ton for every ton of material shipped on the spur. He further proposed that Amco be paid interest on their investment until such time as all costs were reimbursed. Storey still hesitated. Steele finally recommended that AT&SF build the spur at their expense, in return for which Amco would pay them $1.00 for every ton under 250,000 tons of material not shipped during the seven year proposed life of the mine.

This offer aroused Storey's interest, and he offered to send a survey crew to determine a suitable route and more exact cost. However, Steele was reluctant to allow the survey until such time as the mill site and all water rights had been secured. He was fearful that the survey would send a message to the owners of the site and of the water rights that this was a "done-deal," and their property values would suddenly escalate. The meeting was adjourned with an agreement that Norcross would notify Houghton when a survey could be undertaken.

The mill site and all water rights were under option by November 10, 1925 and Norcross notified Houghton to arrange for the survey. He also provided Houghton with an estimate of the tonnage of material which would be received or shipped by Amco should the venture materialize, estimates of which had been developed by Munroe for his report of July 25., 1925 See Fig. 13

ESTIMATED SHIPPING REQUIREMENTS

ITEM	TONNAGE or CARS
Power Plant Equipment	200 Tons
Mill Equipment	830 Tons
Mine Equipment	350 Tons
Tramway Equipment	500 Tons
Cement	10 Cars
Lumber	2.5 M Board Feet
Coal—monthly	1,500 tons
Explosives—monthly	1 Car
Operational Supplies-monthly	7-8 Cars
Zinc Concentrate to Oklahoma—monthly	4,750 Tons (Avg.)
Lead Concentrate to El Paso—monthly	1,350 Tons (Avg.)

Figure 13

A preliminary survey was completed on November 18, 1925; AT&SF concluded that there was merit in expending funds for a more exacting survey and it was finished on December 29, 1925. The route was from Fox to the mill site, a distance of just over six miles. However, they notified Norcross of their decision to survey another route beginning about two and one-half miles east of Fox, a route with a more gentle grade but about two miles longer. This survey would require an additional 14 days before a firm decision could be made by Storey, and Norcross informed Steele of this turn of events.

Steele had expected a firm decision by the year's end. He had instructed Norcross to delay any work at the mill site until that decision was in hand so as to keep the railroad guessing about their intentions. He was angered by the delay and wired Norcross:

> WE BELIEVE RAILROAD WOULD FORCE US TO BUILD SPUR IF THEY SURE WE DEFINITELY COMMITTED MILL CONSTRUCTION STOP SUGGEST YOU FIND OUT FROM HOUGHTON POSSIBLE DATE THEIR FIRM DECISION AND WIRE US YOUR OPINION REGARDING CONSTRUCTION BEFORE RECEIVING RAILROAD DECISION BEARING IN MIND A FEW DAYS MORE DELAY MAY SAVE US ONE HUNDRED FIFTY THOUSAND DOLLARS. STOP SIGNED HEATH STEELE

Norcross wired back four days later:

> SPENT TWO HOURS WITH HOUGHTON TODAY STOP HE STATES: THEY ARE SURVEYING SECOND LINE AND SAID CHIEF ENGINEER HAD ADVISED WOULD BE JANUARY 15 BEFORE ANSWER CAN BE GIVEN STOP MY IMPRESSION THEY ARE STALLING STOP MY OPINION IT MIGHT BE ADVISABLE YOU TAKE INITIATIVE STOP GOODRICH ARRIVES NY TUESDAY AM STOP HAVE GIVEN HIM ALL DETAILS MY VISIT STOP YOU MIGHT BE INTERESTED HEARING FROM HIM BEFORE YOU TAKE ACTION STOP SIGNED NORCROSS

Steele thereupon wrote Houghton the following:

My Dear Mr.Houghton:

As I have already told you several times, our becoming interested in the Pecos Mines in New Mexico was dependent upon the attitude of the Santa Fe Railroad in regard to building the spur at Glorieta. We have just learned—much to our regret—from Norcross that during his talk with you in Chicago on Sunday he found there would be many more days of delay.

We had hoped for your formal decision considerably before this. Consequently we have been placed in a position where the refusal of the owners to consent to further delay on our part has forced our hands. Under these conditions we have decided to go ahead and acquire a controlling interest in the property, relying upon Mr. W. B. Storey's statement to Mr. E. H. Clark—one of our directors—that Santa Fe would build the spur line for us if no great obstacle to the construction were to be encountered.

We are informing you of these facts at once, as it is, of course, our desire to start construction and development at the earliest possible moment. This would naturally be to your advantage just as it would be to ours. We very sincerely hope, therefore, that you will advise us at the earliest possible moment that no great difficulty has been found, and that you are prepared to start work immediately.

Sincerely yours,

Heath Steele

Steele, it is noted, did not mention that fully six weeks earlier, even before AT&SF's survey had begun, a decision had been made to proceed with the venture, the American Metal Company of New Mexico had been incorporated in Delaware, and properties of the Pecos Corporation had been deeded to the new company.

Storey promptly replied to Steele's letter, stating that the route of the spur from Fox to the mill had been selected, and that the cost would be $185,000. He further agreed that AT&SF would build the spur, provided there was a clause in the contract guaranteeing that a reasonable tonnage would be shipped on the spur to justify the construction cost. He stated that Houghton had full authority to deal with Norcross to work out the final details.

Everything ran smoothly until early in February when Haffner took it upon himself to write Houghton a letter demanding to know when AT&SF expected to complete the spur. This is probably the beginning of Haffner's downfall as General Manager. Houghton replied—not to Haffner or to Norcross—but to Dr. Sussman, Amco's Vice President. His letter was marked "Personal," and he began by saying that he did not appreciate the tenor of Haffner's letter. He then went on to say there were too many unknowns this early to give a definitive completion date, but that they would expedite work as much as possible. He pointed out that the necessary rights-of-way had to be secured before any start date could be established. A follow-up letter on February 24 established an estimated completion date of July 15.

Again, all went smoothly until March 30 when AT&SF's on-site engineers advised Norcross that it would be impossible to construct the spur with the mill located as planned, and that the mill would have to be moved downhill to lower the grade of the spur. Otherwise, the cost would increase about $33,000, a sum Amco would be expected to pay. The grade as surveyed was 3.8 percent and the engineers wanted it to not exceed 2.9 percent. Norcross pointed out that a 4 percent grade had often been discussed as a probability in every previous meeting and that none of AT&SF's people had ever objected. He produced minutes of the meetings to prove his point; they backed away from their argument that it was impossible to achieve.

It was pointed out to them that, by slightly increasing the grade up to Alamitos Canyon, and crossing it on a trestle several feet higher, the 4 percent grade they were concerned about could be decreased to 2.9 percent. The engineers argued that there was no locally available fill to increase the approaching grade. When shown where the fill could be obtained, they agreed to consider the matter further. Obviously this was a ploy to get Amco to revise their plans in order to decrease cost of the spur. This would have drastically increased Amco's construction as well as operational costs since, unknown to AT&SF, the tramline contract had already been awarded to Riblet with its terminus as indicated on the mill plans. To move the mill downhill would have required a change in the tramline route from the control station to the terminus. A new survey would have to be made and additional rights-of-way for the line secured. An increase in Riblet's contract price would be an end result.

The problem had arisen because, in initial discussions before mill and tramline plans had been finalized, the original and proposed layout called for the main spur to run below the power house. When plans were finalized, a decision had been made to run the spur line above the power house with a side track into the power house above the coal bins. The reason was a long term economic one; the coal could be off-loaded at a point where it could gravity flow to the power plant's boilers rather than being mechanically lifted from below at considerable cost. The plans had been discussed with the same engineers before being finalized. AT&SF's engineers knew that the change had been made, but they apparently did not realize the significance of the change.

When advised of this turn of events, Steele was willing to give a little since the high-line spur, as it was being called, would result in considerably decreased operational costs as compared to a low-line. He wired Houghton, requesting that there be no delay in continuing with the high-line and promising to reimburse for any additional cost. Houghton wired back that work would continue unabated and cost matters could be settled later. Steele then authorized Norcross to negotiate a reasonable figure for insisting on the high-line route, but only in the event the railroad's estimate of $185,000 was exceeded.

On July 24, 1926, Matson verbally agreed with AT&SF (when Matson assumed responsibility as General Manager of the overall operation, he also took over Norcross' responsibilities) to pay $13,598 of a total $16,007 increase over the $185,000 original estimate. The final agreement on the entire project was to the effect that Amco would pay the high-line cost increase and guarantee to ship not less than 200,000 tons of ore concentrate over the spur by December 31, 1932. Failure to do so would result in a requirement to pay AT&SF one dollar for each ton less than 200,000 tons not shipped during the specified six-year period. As it turned out, the 200,000th ton shipment was made in less than two years and eight months.

Not long after Matson arrived at the mine, he received a late night telephone call from AT&SF's station agent at Glorieta, inquiring as to why he had not received the weekly report on shaft sinking which Haffner had been furnishing. For some reason, Haffner had agreed to supply these figures to the agent on a weekly basis, but Matson could not understand why, neither from Haffner's point of view or that of the railroad. Shaft sinking was only a small part of the total picture and in no way was it an accurate measure of progress being made as to preparation of the mine for production. He declined to furnish the information until he had contacted his superiors.

Steele concurred in his decision and wrote to Houghton to advise him of the decision and the reasons. He offered to have Matson provide their Chicago office periodic reports of the overall progress, but did not want any such information given out at the local level. Houghton replied that he had not been aware that their local agent was getting the weekly reports and readily agreed to receiving such progress reports as Matson would care to provide. He promised to keep the reports confidential. Matson mailed the first on July 15.

The spur was completed by August 26, with no further incidents as to the actual work. However, the legal paperwork was another matter. On July 21, 1926, Matson received from Steele a description of the railroad's right-of-way across Amco's Alamitos property, which had been sent to him for approval by AT&SF's legal department. He asked Matson to check and verify the correctness of the description. The railroad's Resident Engineer advised Matson that there had been so many changes made in the location of the spur that the description would have to be revised. Four months later, Matson was informed by the Resident Engineer that the revised description had been forwarded to the General Engineer in Chicago to rewrite the deed for Amco's approval.

On July 26, 1927, Amco Ltd. received a letter from the General Engineer inquiring as to why he had never received a signed copy of the deed sent to them the previous year. He was in turn apprised of what had occurred and told that the ball had been in their court for almost a year. At about the same time, Matson was handed an original and duplicate copy of the contract for the spur by the railroad's local agent, with a request that they be executed by the Amco Ltd.'s. Secretary. Matson forwarded them to New York for Secretary Hochschild's signature; they were signed and duly returned to the local agent. Two weeks later, a letter was received by Matson from AT&SF's Division Superintendent acknowledging receipt, but objecting to signature by the Amco Ltd.'s. Secretary. He indicated it was customary for the President or Vice-president to sign and be properly attested to by the Secretary, and requested that this be done.

On August 8, Matson wrote to the New York office inquiring as to the status of the documents, only to be advised that the signed documents had been returned on July 5. These, of course, had been the ones signed by the Secretary, not the ones sent for the President/Vice-president's signature, which had apparently been lost in transit. Matson obtained still another original and duplicate which were sent for signature on September 3.

Before these were returned, the revised right-of-way was finally delivered to New York for approval and sent to Matson for his approval. Upon examination by Attorney

Wilson, it was discovered that the revision was so written that Amco would deed the right-of-way in fee simple to AT&SF, with no provision that the property would revert to Amco upon termination of operations. Also, since it virtually bisected Amco's Alamitos property, any necessary changes in pipelines or other structures crossing the right-of-way would require prior approval from the railroad. Matson advised Steele of his objections, and on January 24, 1928, wrote to the railroad's Chief Engineer requesting a revision to eliminate the fee simple deed in favor of a lease for the right of way. By mid-February, all documents were corrected to the satisfaction of all concerned, and the formal contract was signed by Steele and attested to by Hochschild on February 18, 1928.

Construction of the spur had begun in March, 1926, had been completed in August, 1926, and had been used for freight traffic until February, 1928, without benefit of a formally signed contract. It had all been accomplished because the presidents of the two companies had gained mutual respect for each other's integrity. Steele and Storey had begun by distrusting the other's motives; they ended the venture good friends and with high regard for one another.

chapter 10

* *

Operational Potpourri

Because Haffner had purchased numerous expensive items not directly required for mine development and production, a decision had been made that any such items purchased in the future would require Steele's pre-approval. On June 13, 1927, Matson submitted authorization request #1 to purchase a new five-passenger Studebaker sedan at a cost of $1,900. In the letter he pointed out that, during the past six months $900 had been expended for transportation of visiting dignitaries from New York or other Amco Ltd. operations. In his reply, Steele authorized the purchase, but questioned the wisdom of purchasing a Studebaker instead of a Buick and stated, "In the writer's opinion the Buick is by far the best car on the market for the money."

Matson replied as follows:

> I note in your approval of our request to buy a car that you wonder why we had chosen a Studebaker. The principal reason for this is that the Studebaker has a standard gear shift and is therefore more readily adaptable for anyone needing to drive it. Brown in particular dislikes driving any car without a standard gear shift and one with a standard shift would therefore be more convenient for him. I have driven a number of makes of cars around here on trial and agree that the Buick is a good car for the money. However, there are some objections to it. I found that after driving one a few miles the brakes are not thoroughly reliable. This is particularly so in trying to put the brakes on when backing up. In these mountains that could be disastrous. Also they seem to lack much power in high gear. While the Studebaker may have these same problems in the long run, I have not proven it so far. If it does, I will soon know for we have purchased one. The prices of the two cars were practically the same.

No evidence could be found that other authorizations were ever requested.

During the first few of years of operation, Matson was continuously being asked to find a positions for one person or another—for some friend of a top Amco Ltd. official or of another general manager of an Amco Ltd. operation, or of some governmental or educational entity. In the first year alone, requests for nine different individuals were received. Only three were honored—those that had been received from Steele. They were Jim Welsh, the postmaster; Hans, Steele's nephew who wanted a job while on college summer vacation; and a fellow named Jack Finlay. The nephew was put to work as an office boy, but Finlay had no particular work experience. Steele suggested he might be used in the engineering department, but he was eventually hired for a position in the accounting department. Such temporary employment would lead to two tragic deaths.

In late 1928, Osborne C. Wood, Adjutant General of the New Mexico National Guard, came to Matson with a special request for himself. During World War I, Matson, a captain, had served as adjutant to Osborne's father, General Leonard Wood for whom Fort Leonard Wood, Missouri, is named. During that time he had met Osborne who was serving in France as an army lieutenant colonel. The father had died in August, 1927, and had stipulated in his will that his son could inherit his sizable wealth on one condition. He had concluded that Osborne had life too easy and that he should have to work at some form of hard labor for a year before he could receive his inheritance.

Having learned that Matson was General Manager of the mine, Osborne came to him, told him the story of his father's will, and requested that he be hired as a mucker for one year, the hardest of all mine work. Matson obliged. Wood requested and was granted a one-year leave-of-absence from the National Guard during 1929, worked the year as a mucker, and returned to the Guard as Adjutant General on January 1, 1930.

During the summer of 1931, he flew over Tererro in a National Guard airplane and took the frontispiece aerial photo at the beginning of this book. He gave the photo to Mr. Matson, who left it to his son Joe, who in turn left it to me. Joe and I, as well as many others, had watched in fascination and conjectured as to why the National Guard plane circled above Tererro at about 12,000 feet that summer morning—flights during the afternoons would have been almost impossible due to the frequent afternoon thunderstorms. Very few ever knew the reason the flight had been made.

During the summer of 1928, another student from the New Mexico School of Mines was temporarily hired at the request of his brother, the State Geologist for New Mexico. Around 2 A. M. on August 7, the young man, Lyle Staley, was on the lift cage returning to the surface with five other men. After they entered the cage, it had moved up only about 15 feet when he suddenly fell forward and onto the floor. Although the cage was fitted with a steel gate, there was room between the bottom of the gate and the floor of the cage for his body to pass through. His leg apparently slid out into the small space between the cage and the shaft lagging, caught on something and his body was pulled through. He dropped 300 feet down the shaft to his death.

All five men riding with him said that he made no sound as he fell forward or as he fell to his death, so Dr. Smith's autopsy concluded that he had fainted and never knew what happened. The investigation could not ascertain why the other five men could not have kept him from passing through the opening, unless his leg had in fact caught and he was pulled through as the cage moved upward.

His was the second fatality of the year. A miner, S. L. Hern, had been the first since operations began, having been killed in a cave-in a few months earlier. The liability of the company's insurer, the Maryland Casualty Company, was $75, the maximum permitted by law at that time. No relatives or close friends could be located, so a decision was made that Amco was morally obligated to provide him a proper burial in a Santa Fe cemetery. They paid an additional $89 for the funeral. Another cave-in later in the year resulted in four more deaths, for a total of six during the year.

No records of accidents or safety problems had been kept up to the time of the four deaths, and no record could be found as to their identity. However, those deaths resulted in a decision to hire someone to establish a safety program. Appendix D summarizes the accident records which were kept beginning January 1, 1929.

When Norcross was looking at properties to buy just after the Amco Ltd. decision was made to purchase the 51 percent interest in the Pecos Corporation, he spoke to a carpenter who had been involved in building the Simmons ranch facilities, now used as the hospital and residence for Dr. Smith and his family. He showed Norcross several pieces of ore that had been excavated when a dam across Holy Ghost Creek had been built to create a small pond in front of the house.

Norcross wrote to Dr. Sussman the following with regard to the ore: "I believe we are absolutely safe as to the knowledge of possible ore on the Simmons' place is limited to myself and a half-witted carpenter who put in the Dam (sic) and brought me several pieces of the ore he dug out and I told him it was 'good road rock.'" In the same report, he states that outcrops were said to lie under the Simmons' old ranch house.

He kept the ore samples given to him by the carpenter, had them assayed, and filed a lode claim for the company on the potentially valuable body of ore located beneath the dam. Also, it was presumed that if there were ore bodies at both Willow Creek and Holy Ghost junctions with the Pecos, there might also be ore deposits between the two.

In December, 1929, a diamond drill rig was set up at the site of the hospital dam and two holes were drilled, one angled 1,300 feet deep to the west, the other 400 feet deep to the east. Both proved to be without result. A trench was dug across Holy Ghost Creek, parallel to and directly below the dam, to find the outcropping reported by the carpenter. It too proved valueless, and the venture was discontinued.

Diamond drilling within the mine, however, was much more successful. A couple of new, smaller ore bodies were located, and the Evangeline and Katy Did were found to go much deeper than had previously been determined. Extensive mineralization had been located by early 1931 at the 1,000 level and, in 1933, at the 1,200 level. The original estimated seven-year life of the mine had been extended by at least three more years.

On July 1, 1927, the Forest Service relinquished all claim to the Pecos River road between Pecos and the mine. In attempts to get San Miguel County to assume responsibility for its upkeep, Amco was unsuccessful. However, the County did agree to pay the company $4,000 for road improvements the company had previously made, but only $1,000 had been received and it seemed doubtful more would be forthcoming during the remainder of the year. As a result, Matson decided to approach the State Highway Department to see if they would assume any of the upkeep costs. As a result, the Highway Department agreed to assume responsibility for the road's maintenance at a cost to the company of $3,500 per year for four years.

It had become clear early in the year that Dockwiller could not meet the timber needs of the mine, so a timber cutting-permit was obtained from the Forest Service. The cutting area was established on the eastern slope of the Willow Creek watershed. After cutting, the ponderosa pine trees were trimmed, peeled, and cut into 16-foot lengths. The timber averaged 5 to 14 inches top diameter and was hauled on team-drawn skids to a timber yard near the collar of the Evangeline shaft.

Equipment there consisted of a swing cut-off saw, a rip saw, and a framing machine. The smaller poles, averaging about 6 inches in diameter, were used for round raise cribbing, and any oversupply was split into half-round lagging. The larger timber, from 8 to 14 inches in diameter, were framed into square-set or drift timber. Additional lagging was obtained by slabbing the larger poles into two-inch plank with round edges. Unusable ends were cut into firewood and sold to residents at $2.00 per cord, so the only waste was the sawdust. The cost of all timber thus produced was estimated to be about $16 per thousand board feet.

By the end of 1927, the supply of timber in the Willow Creek area became depleted and the Forest Service designated an area on the slopes of upper Davis Creek for further timbering operations. Five miles of road were built into the area. About two miles from the mine, a sawmill was set up at the 9,800 foot level near the head of Davis Creek. Dockwiller was hired as the operator. Two large Caterpillar tractors and several tractor trailers were purchased to replace the horse-drawn skids. The cost of timbers delivered to the mine yard was increased to about $20 per thousand board feet, however the amortization costs of the mill and handling equipment was included in the total. At this figure, approximately $5,000 worth of timber was being consumed each month.

By 1933, the timber in the upper Davis Creek area had also been depleted and the Forest Service issued a permit to cut timber in Indian Creek Canyon. The sawmill was moved to a site about one mile up the canyon from Irvin's Ranch, and a small community of houses was built there by those who would be employed in the operation. Ralph Littrell saw an opportunity to better himself financially, and presented Amco with a bid for hauling timber to and from the sawmill. His proposal was accepted.

With a contract in hand, he was able to purchase four trucks and logging trailers, and the materials to build a house at the Indian Creek site. He operated the enterprise until 1937, when he sold it to one of his employees. Eventually he ended up working as a blacksmith's helper at the mine.

San Miguel County's assessed value of the mining and milling operation was $511,000 in 1926, including an assessed value of $1,000 placed on each of 33 mining claims. Taxes for the year were $12,775 at a tax rate of $25 per thousand. In 1927, the assessed valuation was decreased to $464,257, but the tax rate was increased to $35 per thousand, for a total tax of $16,249. Based on increased production in 1928, it had been anticipated that the valuation would go to $1,000,000, and that the rate would increase to $45. Through negotiation with the tax commission, however, the increased valuation was placed at $750,000, but the rate was established at $29, for a total tax of $22,000.

During the negotiations to establish the assessed value, one of the tax commissioners stated, "I consider the Amco operation more or less a public enterprise in that the indirect benefit of your operation to both State and County is of more importance than the collection of taxes." He obviously was referring to the mine's very positive impact on the economy—the employment at that time of about 400, the approximately $50,000 per month payroll, and the mine's local purchases of approximately $70,000 in goods and services each month. Since this was after only one year of operations, the figures could well be expected to double in the future.

Mr. Seligman, president of the First National Bank of Santa Fe, indicated that 1927 was the most prosperous year in the history of Santa Fe and credited the Pecos Mine's business and payroll as being responsible for the exceptional business conditions. A 1927 aggregate of $117,600 in purchases was awarded to ten Santa Fe companies, the Santa Fe Builders Supply Company alone having received $55,200.

After the end of the first year of operation, sales of zinc and lead concentrates had amounted to $1,959,773.40, while the cost of operations was $1,209,069.97—a total gross profit of $750,703.43. In 1928, gross profits rose to $846,689.73 and, in 1929, to $1172,254.14. The stock market crash of 1929 brought these high profits to an abrupt end. Total gross profits for these first three years exceeded those in the last ten years of operation by almost double. See Appendix C

Two types of Federal law violations plagued the company early in its operations. Numerous bootlegging operations were springing up, particularly in the Hispanic sector of the mine camp. On paydays, the 5th and 20th of each month, a considerable number of the employees would go on bootleg-whiskey binges and be off work until they sobered up several days later. In May of 1927, 209 man-shifts of work over seven working days were lost to this drunkenness.

Efforts to obtain assistance from both the Federal government and local authorities proved to be fruitless. The attitude seemed to be that this was too small-scale for the enforcement agencies to bother with. Matson wrote to Steele, "The only action for us to take is apparently a thoroughly drastic one regardless of the law. Otherwise, we will soon have a wide open camp and a resulting impossible situation." There was never any further mention of the problem in any of his future reports to Steele, so whatever action he took apparently was successful.

The other problem had to do with a complaint from the El Paso office of the Federal Immigration Office that the company was recruiting and hiring illegal Mexican nationals. This no doubt stemmed from the fact that the company's contracted recruiter was perhaps actually recruiting in the El Paso area and picking up some illegal aliens in the process. Matson agreed to do what he could to have the background of all Hispanic employees investigated, and if any were found of suspect nationality, to notify the Albuquerque office of the Immigration Service. This problem was never mentioned again in any correspondence, so it can be assumed that it, too, was taken care of satisfactorily.

chapter 11

* *

Concerns for the Environment

Early in 1929, at the behest of Ed Irvin, owner and operator of Irvin Ranch, located at the juncture of the Pecos River and Indian Creek, a bill was introduced in New Mexico's House of Representatives to make it illegal for any mining company to "....cause to be emptied, or allow the emptying or blowing of any sawdust, tailings, oil, petroleum, refuse, slops, or other obnoxious or poisonous matter, or any silt or other matter pumped or taken out of any mine or mines, or in any other manner to pollute the waters of any stream, or other public waters of this State, or to deposit or leave any of such substance or other substances where the same may be carried by natural causes, or otherwise, into any such stream, or other public waters in this State." Penalty for conviction of violation of the provisions of the law was to be set at not less than $50, nor more than $300, for each day that the pollution continued after such conviction. The bill passed in the House and was sent to the Senate.

In the general election of 1928, Dr. Smith was elected, by almost a 2,000-vote majority, to the New Mexico State Senate. During the 60-day legislative sessions from mid-January to mid-March each year, a Dr. Williams, formerly with the State Health Department, substituted as the company doctor. Dr. Smith agreed that he would take leave from any further legislative work should a serious medical emergency occur.

When the water pollution bill was debated in Senate Committee, Dr. Smith opposed it vigorously, and it died there without having become law when the Legislature adjourned at the end of its term. Mr. Irvin, however, did not give up. He complained to the State Game Commission that silt from the water being pumped out of the mine was killing fish in the river and seriously affecting his dude ranch operation.

To put the matter to rest, the State Game Warden requested Matson to have a test pond constructed at the mine, to be fed only with water pumped directly from the mine. A ten-foot by ten-foot by four-foot in depth concrete pond was thus constructed below the mine office, and was stocked on May 15, 1930, with 209 six- to ten-inch trout. Screens were placed at both the inlet and outlets to prevent escape of the fish. The State Fish Hatchery provided food that was fed to them by "Shorty" Gallegos, the office janitor and Matson's handyman.

During the following six months, 21 of the trout died and were removed from the pond by "Shorty." On October 15, 188 of the original trout were removed, at which time they measured from 8 to 14 inches in length. In addition, 42 smaller fish, that had passed through the outlet screen from Willow Creek, were also found in the pond, some having grown large enough that they too could no longer escape. In the bottom of the pond, over 17 inches of mine silt had settled out. The only deleterious effect on the fish had been a slight change in their coloring which was deemed normal, having been caused by the fact the fish were constantly in water made opaque by the silt in suspension. The silted water, as it flowed from the mine, was not too unlike that which flows from beneath some glaciers in Canada and Alaska.

In his report to the State Game Commission, the State Game Warden wrote:

> This experiment proves conclusively that there is not a sufficient amount of deleterious silt or other material in this water to affect the fish in the Pecos River in any way. The volume of water in the Pecos river is so great that this small stream emptying into it would have less effect there than it would on the fish in the pond. The fish in the pond were able to survive in good shape throughout the season with only a normal loss. The percentage of loss was no greater than is found in the fish hatcheries using the best possible supply of water.

During the summer of 1929, a large test garden plot was planted below the tailings dam across Alamitos Canyon which was irrigated totally with dam overflow water. Garden produce from this test plot was entered in the San Miguel County fair. Three first awards, along with several second and third place awards, were won, indicating that overflow water had no damaging effect on the growing crops.

However, in the first week of September, 1931, the County Commissioners of Guadalupe County (the county bordering San Miguel County on the south) held a meeting in Anton Chico, a small village on the border between the two counties, to discuss with the farmers along the Pecos River that summer's failed chili crop. The farmers placed the blame on the mill water that emptied into the Pecos, from whence they drew irrigation water. Bemis and several others from the mill attended and spoke at the meeting A Dr. Garcia from the New Mexico State Agricultural College was also present and proved to be the key voice in settling the controversy.

Bemis declared that the water from the tailing pond was carefully monitored to insure that damaging chemicals did not enter the Pecos. He voiced the opinion that the chili crop failure could not have been caused by the mill and was sustained in this contention by Dr. Garcia. Dr. Garcia testified that there was no proof to support the farmer's complaints; rather, he stated that there was adequate proof to the contrary.

He introduced a San Miguel County farmer named Griego who had irrigated his chili crop extensively with Pecos River water for many years. Griego said that he had never had a crop failure either before or after milling operations began, and that the past year's chili crop was one of his best. It was pointed out that his farm was about twenty miles closer to where Alamitos Canyon joins the Pecos and would probably have had a greater problem with the water if it was, in fact, polluted with the mill water.

Dr. Garcia urged the farmers to look elsewhere for the cause of their failure, since he believed they could not prove damage from the mill. This put an end to that particular problem, but Matson issued instructions that the mill be even more vigilant in insuring that there be no pollution attributable to the mill. He proposed in writing to Steele that a new tailings dam be constructed further downstream to avoid any possibility of a heavy rainstorm causing undue overflow and the washing of tailings into the Pecos. Vice-President Van Dyne Howbert was sent out from New York to personally look at the situation, and he fully concurred in Matson's recommendation. The new dam was designed, let out to contract, and constructed in record time.

On occasion, when someone would question whether or not the water flowing from the tailings pond was perhaps polluted with chemicals from the mill, certain employees would delight in dipping out a cup of the water and drinking it down. Dr. Smith was one of the first to do so. At that time, this seemed to be proof enough that the water was not harmful. This is not to say, however, that there may not have been some long-term effect on one's health had the undiluted water been consumed on a day-to-day basis.

Despite Amco's efforts to absolutely insure that the mining operation did not damage the environment, changing knowledge and technology would, over the fifty years following the cessation of operations, call into question whether or not due care had been used in protecting the environment from the effects of lead poisoning. This aspect of the venture will be further discussed in the epilogue.

chapter 12

✲ ✲

A Futile Search

From the outset of mining operations, gold had proved to be an unexpected income source, particularly so in 1929. During the early years of the depression, when gross profits had decreased dramatically, See Appendix C. efforts were intensified toward finding additional gold and other mineral resources to help bolster the declining revenues. This became paramount when the gold standard increased to $35 per ounce on January 31, 1934—an almost 70 percent increase in its value. One potential source of additional revenue lay in increasing the amount of recovery of copper and gold as compared to that which was currently being extracted in the milling operation.

Another source lay in recovering some of the gold and silver which had already been lost to the mill tailings. The difference between the assayed value of gold at the head of the milling process and the value of the recovered gold after milling indicated that at least 20 percent was being lost to the tailings. If any of these three minerals—gold, silver, and copper—could be profitably recovered, the profit picture would be attractively enhanced.

In early 1932, Professor A. J. Weinig at the Colorado School Of Mines was engaged to conduct tests on the lead concentrate, as it was then produced at the mill, to ascertain whether or not an additional copper concentrate could be separated out while simultaneously separating out additional quantities of gold. His test report is dated April 22, 1932.

Through the use of a high-level hydrated lime circuit and a reagent known as Minerec that floated the lead away from the copper, he reported only limited success in getting a good copper concentrate, but a fairly substantial additional recovery of gold. By adding cyanide to the circuit, a much better copper concentrate was obtained, but the gold was depressed into the tailing. The probability was high that if this process were to be added to the milling process, there would be an increase in the amount of gold lost to the tailings. This loss would not be offset by the increase in copper recovery.

In a third test, cyanide was replaced with a reagent called Ceraline. The separation of a moderately good copper concentrate was achieved, although not equal to that with cyanide as the reagent. The separation of additional gold was about equivalent to that obtained in the first test using Minerec without the cyanide.

Weinig recommended use of the third process if it could be determined that it would present an economic advantage over the then milling practice. An economic analysis indicated that, when the cost of additional equipment was added to the cost of the hydrated lime and Ceraline, little would be gained financially; and the idea was discarded. Two years later, the gold standard had increased to $35 per ounce and Weinig was engaged to test another potentially profitable idea.

Assays of samples taken from the tailings pond indicated that the tailings contained roughly .022 ounces of gold and .7 ounces of silver per ton. The tailing pond at that time contained approximately 1.25 million tons—a potential of almost $1,000,000 worth of gold and $400,000 worth of silver. If a reasonably large percentage of this could be recovered using some simple recovery process, and if an additional percentage of the gold were to be removed in the initial milling process, profits could be increased dramatically.

The tailings were composed of approximately 50 percent, by weight, of material classified as sand (100-mesh or larger), while the remainder was classified as fines (between 100- and 200-mesh) and slimes (200-mesh or smaller). Weinig's efforts were concentrated on recovery from the sand, since less than 20 percent of the tailing gold was found in the slimes; however, control tests were also conducted on the fines to insure that no potential revenue enhancement had been overlooked. The sands assayed at ± .035 ounces per ton, the fines at ± .015 and the slimes at ± .008 ounces. The silver was equally distributed throughout the tailings.

Weinig's final report was issued on October 18. Nine samples had been collected from the tailings pond at strategic locations. For testing purposes, samples 1, 2, and 3 were combined, samples 4, 5, and 6 were combined, samples 7 and 8 were combined, and sample 9 was used as collected. Each of these four combinations were separated into two batches, or heads, one classified as sands and fines, the other as sands and slimes. These heads were then separated into their respective components (100-mesh separation for sands and fines, and 200-mesh separation for sands and slimes). The average assay values of the heads (as shown in the previous paragraph) were then determined, although there was considerable disparity from one head to another.

Tests were then conducted on the sands from both separations by varying the temperature, the concentration of the cyanide solution, the addition of mercury chloride, the amount of agitation, and regrinding of sands to obtain a slime so as to decrease processing time. A total of 20 different tests were conducted, and in one test (the least expensive overall process) 75 percent of the gold was recovered and 33 percent of the silver. Unfortunately, this test could never be duplicated. In another test (the most expensive overall process) 82 percent of the gold and 31 percent of the silver was recovered. The extreme cost of this process definitely precluded its use as a production process despite the improved recovery rate.

In his covering letter for the final report, Weinig had this to say, "These tests will probably not figure economically, but you may feel they offer enough encouragement to do some further work along this line. I shall be glad to have you advise me in this regard. There is this to be said in favor of a concentration scheme if anything worthwhile could be developed: The existing mill might be employed after the mine ore is exhausted and thus avoid the high capital expenditure such as will be required to install equipment for cyaniding the tailings." An economic analysis showed that it was not feasible to continue further investigation, and the matter was dropped. [1]

In mid-1935, a body of ore was found in drift 1182 (located at the 1100 level) which assayed extremely high grade gold/silver values. This was the first time this kind of ore had been seen in the mine, and it evoked considerable interest. Mr. Charles Stott, Chief Mining Engineer at Amco Ltd.'s Presidio silver mine in Shafter, Texas, an expert on high grade gold/silver ores, was dispatched by Steele to Tererro to evaluate the find. In his report back to Steele, he states,

> The gold/silver values in 1182 drift are definitely associated with a lead-gray sulphide, which is probably argentite. This mineral occurs elsewhere in the mine associated with lead-copper sulphides. However, in the 1182 drift it is not associated with the normal Pecos sulphides, but occurs isolated by itself. As no free gold is seen in hand specimens, or by the microscope, the gold values must necessarily be bound with the argentite in variable proportions.
>
> Specimen samples were taken across the entire showing for a width of eight feet and the results correspond roughly with the cut samples and the diamond drill hole. The specimen samples assayed between .22 ounces gold, 4.68 ounces silver and 5.18 ounces gold and 38.6 ounces silver within this mineralized width. Sixty tons of muck mined from fifteen feet on the drift averaged .84 ounces gold and 10.0 ounces silver.

Stott went on to explain in great detail that a review of the records indicated argentite (silver sulphide, Ag_2S)bodies had been found in other drifts (sans the gold/silver values) and that the occurrences were always in shear zones. He further explained that their shape and habit were identical with the normal sulphide ore bodies found elsewhere in the mine. He laid out a development program to seek out similar ore bodies along the main shear zones of the mine, and he detailed what kind of and types of rock should be an indication of nearby argentite bodies should they be found in diamond drill cores. He also suggested a new look at all cores previously taken from unmined

---

[1]. In 1974, the Natural Resources Development Company of Santa Fe made a technical proposal to the New Mexico Department of Game and Fish (current owner of the mill property) proposing to establish a gold recovery plant at the tailings ponds (two were eventually filled). They estimated that there was, at that time, $18,000,000 in recoverable gold. They offered to construct a pilot plant operation to determine the feasibility of their proposed process, which they described as thermal-pneumatic concentration. The pilot plant was to operate for a period of no more than six months. If proven to be economically sound, a production plant would be constructed and all gold recovered within a period of five years. They offered the Department a royalty payment of 5 percent of the gross profit. The estimated value of the gold was based on $100± as the price of gold prevailing at that time, about three times the value of gold when the mine closed, but as the report also stated the cost of recovery had increased about three times. The proposal was not acted upon by the Department.

In 1986 and 1987, Amax, the successor company to Amco Ltd., again evaluated the feasibility of extracting precious metals from the tailings in connection with planned reclamation work (see Epilogue). Even with the then prevailing gold price of around $400 per ounce, recovery was found to be uneconomical.

areas to see if these indicators might be present. He concluded his report with the following:

> Although the ground to the southwest and directly below 1182 has been well prospected and sampled, with disappointing results, the heavy occurrence of argentite in a good shear zone such as 1182 is very encouraging and leads me to believe other ore bodies of this type will be found as development is continued.

As a result of Stott's report, and because this particular ore body proved to be the richest find to that date, a great deal of optimism was generated at the management level. The corporate office in New York dispatched a geologist, Philip Krieger, to conduct further tests, and to confirm Stott's findings. His report was completed in mid-November and his findings were almost identical to those of Stott, as were his recommendations.

It was hoped that similar occurrences of this high value gold/silver ore could be located, and would be the added income resource sought after during the past few years. Despite a vigorous diamond drilling program, it was not to be—no similar ore bodies could be located. Diamond drill prospecting was ended in December, 1937. This was the beginning of the end of the mine. On February 12, 1936, another event occurred which would have an equally vital bearing on the ultimate decision to cease operations.

chapter 13

* *

Strike!

Around seven on the morning of February 12, 1936, the local peace officer employed by Amco, a Mr. Miller, phoned Matson at home to tell him that a large number of men had assembled at the collars of both shafts and had prevented the day shift pump operators from going down the shafts to their work stations. The avowed purpose was to cause the mine to flood if Matson did not agree to re-hire a man who had been discharged for cause several days earlier. The men were led by Andres Cruz, President of Local 64 of the International Union of Mine, Mill, and Smelter Workers, Harry Gennis (also known as Harry the Greek), and Adolpho Maes. All three men had been terminated for cause during the past several weeks, Cruz having been discharged on February 1, and elected as president of the union four days later. Harry the Greek had been involved in strikes at coal mines in Gallup, Raton, Dawson, and Madrid during the two previous years. Matson, in his report of the incident to Steele, refers to him as a well-known member of the Communist Party and a radical agitator.

Miller was told to return to the mine to warn the men that they were committing an unlawful act and would be subject to arrest for trespass if they did not disperse (the mine was not unionized and had no contract with any employee organization). Matson then telephoned the County Sheriff and the State Police for assistance in dealing with the men. He then went to his office, called Francis Wilson, the company's attorney in Santa Fe, and called his staff together to discuss a plan of action.

Superintendent Cliff Hoag and Master Mechanic Ole Lee were sent to enter the mine and request the graveyard pump-men to remain on the job until the matter could be resolved, or the day shift pump-men could be brought down. The strikers, meanwhile, sent a committee of four to the mine office to verbally present Matson with their demand, namely that Cruz be reinstated. Otherwise the strikers were prepared to stop the pumps by physically preventing pump-men from reporting to work, thereby flooding the mine. At that time, they held the upper hand. To stall for time until the Sheriff's Deputies and State Troopers arrived, Matson had kept them cooling their heels in the outer office. It was absolutely essential to keep the pumps working and to avoid any confrontation with the strikers until there was an adequate law enforcement force present to deal with the uprising. The County Sheriff arrived with nine deputies about ten a.m., and a captain of the State Police arrived with six officers soon thereafter.

Matson briefed the Sheriff and the Captain, who then took their men up to the mine and ordered the strikers to disperse. They refused; several tear gas canisters were lobbed into the crowd and the area around the mine was quickly cleared. Meanwhile, Matson met with the committee of four and asked them to submit their demand in writing. The committee left and returned about one-half hour later with the following written demand which I have copied verbatim:[1]

Tererro, N./M., Feb.12,1936

Following are the demands that Local #64, International Union of Mine, Mill, and Smelter Workers, officially has decided to present to Mr.J.T.Matson,General Manager of the American Metal Co. of New Mexico.

1 We want the Union to be recognized.

2. We want our leaders Andy Cruz and Harry Gennis to be put back to work, also Adolfo Maes and Daniel Hererra.

3. We want $1:50 raise per day

4. Installation Sanitary System in the camp and mine.

5. We demand fair prices and better quility at Henry Pick's store and abolition of the "Scrip System",and cupon business.

(Signed) Andy Cruz

Andy Cruz,President,Local #64,
International Union of Mine Mill and Smelter Workers,
Tererro,N.M.

(Seal Reading)

TERRERO MINE
WORKERS UNION
NO. 64, ORGAN-
IZED AUG 8, 1934
I.U.OFM.M.&S.W.

Although, with the mine entrance cleared of strikers, it would have been possible to get a crew of men who were not involved in the strike down in the mine to resume work, a decision was made to operate only the tram until the mine ore bin was emptied, and to continue to operate the pumps and the hoist to move the pump-men in and out of the mine.

Later in the day, the Sheriff and the State Police Captain visited the Union Hall (an old store building located on the adjacent Louis Rivera property) and warned all who had gathered there that any trespass on company property in the mine complex itself was prohibited, either by individuals or by groups of individuals. There were shouts that they would do as they pleased. About an hour later a group of around 400 shouting men marched toward the mine, but were stopped at the bottom of the steps leading up to the mine by the two law enforcement agencies. Both Cruz and Gennis were allowed to stand on the first landing and speak to the assembled strikers advising them that they had been warned by the law enforcement people that they would be placed under arrest if they attempted to enter the mine complex.

Attorney Wilson arrived from Santa Fe around two p.m. He and Matson met with the Sheriff and State Police Captain to discuss the next step. The Captain advised that the State Police were also involved in a strike at the Madrid coal mine now in its third week. He suggested that based on their experience at Madrid, for at least several days, no work should be done except to keep the pumps operating. Both the Sheriff and the Captain

[1] This document is copied precisely as it was written including all spelling, punctuation, spacing, and grammatical errors.

identified the leaders of the strike, except Cruz and Maes, as avowed Communist agitators who had led a strike at Gallup the year before. In that strike, the County Sheriff was murdered and martial law was declared in Gallup by the Governor. The leaders were Harry Gennis, Roberto Palleros (under orders of deportation as an alien Communist), Fred Arrellino, and a lawyer from Arizona named Lynch. The two Hispanics belonged to a Communist organization known as Ligue Obrera which became involved in each of the strikes before-mentioned.

That afternoon, difficulty was encountered in getting the second shift pump-men to report for work; it was learned that a group of four strikers had visited the homes of these men and threatened physical harm to their families if they went to work. At the times of the visits, dynamite was exploded near the homes, showering them with rocks and debris, obviously an attempt at intimidation.

Attorney Wilson spent the night in the Matson guest house. The following morning, with considerable evidence in hand, he went to Las Vegas to consult with the District Attorney, who agreed that the evidence warranted criminal proceedings. Warrants were issued and the four, Gennis, Cruz, Babb, and Henderson, were arrested on complaint of the men whose families were threatened. Gennis and Cruz were additionally charged with inciting to riot and attempted destruction of property on complaint of the Sheriff. All but Gennis were able to post bond of $5,500 each and were released.

On the 14th, a decision was made to shut down operations indefinitely, keeping only the hoist-men and the pump-men on the job. The following notice was posted throughout the town.

NOTICE

THE AMERICAN METAL COMPANY OF NEW MEXICO, FOR OVER A YEAR, HAS BEEN IN LOW GRADE ORE AND HAS HAD TO CONTEND WITH INCREASING OPERATING COSTS. WHETHER IT SHOULD CONTINUE OPERATIONS HAS BEEN FREQUENTLY DISCUSSED BY THE OWNERS. THE LARGE NUMBER OF EMPLOYEES AND THEIR DEPENDENTS HAS BEEN ONE OF THE IMPORTANT CONSIDERATIONS WHICH DURING THE PAST FIVE YEARS HAS CONVINCED THE COMPANY THAT IT SHOULD CONTINUE OPERATIONS.

IN VIEW OF THE PRESENT DISTURBED CONDITIONS AND THE ONGOING STRIKE, BROUGHT ABOUT BY OUTSIDE AGITATORS AND CONDUCTED UNDER THE AUSPICES OF LOCAL 64, INTERNATIONAL UNION OF MINE, MILL, AND SMELTER WORKERS, THE ONLY RECOURSE LEFT OPEN TO THE COMPANY IS TO DISCONTINUE PRODUCTION OPERATIONS INDEFINITELY.

THE MINE WILL BE KEPT DRY AND SOME DEVELOPMENT WORK WILL BE DONE IN THE HOPE THAT MORE ORE WILL BE FOUND WHICH WILL ENABLE THE COMPANY TO RENEW OPERATIONS SOME TIME IN THE FUTURE. IF THE DEVELOPMENT DOES NOT SHOW SUFFICIENT ORE TO WARRANT RESUMPTION, THE PECOS MINES WILL CEASE TO EXIST.

THE AMERICAN METAL COMPANY OF NEW MEXICO,

By; (Signed) J. T. Matson
General Manager

Dated at Tererro, New Mexico
4 p.m., February 14, 1936

The tram and power lines, as well as the bridges across the Pecos River, were extremely vulnerable to being sabotaged if there was any attempt to keep the mine operating, and it would be impossible to guard against such outcome. With the known record of the strike leaders, excepting Cruz and Maes, who seemed to be pawns of the others, the possibility of such sabotage was an almost certainty. Despite the notice, further development work would be postponed until the situation had stabilized.

For their protection, all of the families of the hoist-men and pump-men were moved to Las Vegas or Santa Fe at company expense. Thirty-five guards to secure all vital installations were hand-picked from those men who had not joined in the strike and wanted to work. In addition, the Sheriff's Chief Deputy and several other deputies were on duty as were four State Police officers. The salaries of all except the Chief Deputy were paid by the company. The four State Police officers monitored the picket lines which assembled at 7 a.m., 3 p.m., and 11 p.m.

Initially, about 400 men, and sometimes their wives, would set up picket lines at three places along the highway—at the bottom of the stairs leading to the mine, at the juncture of the road leading from the highway up to the mine, and at the first tramline. shed. As the days went by, the number dwindled to about 165 at three p.m. and 100 on each of the other two shifts. For the most part, the same men picketed at all three times, that is, there were not three different groups of pickets.

The Ligue Obrera organization began passing out pamphlets on the street corners of the Santa Fe Plaza, castigating the managements at both Tererro and Madrid for not capitulating to the demands of the two striking groups. They further solicited funds from both stores and individuals in Las Vegas and Santa Fe to help feed the families of the strikers. Editorials in the *Santa Fe New Mexican* and the *Las Vegas Optic* were objective and defined the issues at both mines quite well. They both suggested that the strikers might be better served to bargain in faith for a settlement of the strikes. The Tererro group was adamant—they would not give in without full acceptance of their demands by Amco's management. The management was equally resistant and refused to talk with the strikers until, and only if, they were willing to concede on some points. The parties were at a stalemate.

Matson made the decision to begin additional development work on March 1. About 100 non-striking employees were asked to return to work, including 19 residing in Pecos. For mutual protection, those 19 attempted to drive to the mine in a convoy, but a considerable group of strikers, having gotten wind of their plan, set upon the convoy just north of Valley Ranch, stoned their cars and attempted to prevent them from proceeding. One of the men, a night foreman, was injured by flying glass from his smashed windshield, and his car was forced off the road and severely damaged.

Some of the non-strikers carried guns and two of the strikers were shot in the melee; they were not seriously injured, having received only some buckshot pellets in their legs. Fortunately, a group of State Police happened along about that time and arrested 15 strikers who were then incarcerated in Las Vegas. The non-strikers resumed their trip to Tererro and reported for work about an hour late.

A few nights later, dynamite was used in an attempt to blow up the water storage tank above the mine; fortunately, only minor damage occurred. Had it been successful, the town would have been without a piped water supply, although water could have been hauled from the Pecos River or Willow Creek. Five men were arrested for the incident and Juan Rodriguez who led the group, made a written confession implicating Cruz as the mastermind. He also implicated the four others arrested with him. Cruz was re-arrested on an additional charge of attempted destruction of private property.

After these arrests, things began to quiet down, but the stalemate continued and no meetings were held between the disputants for several days. Matson's biggest concern was that many of the best men, particularly those who were mine shop-men and mechanics, were among the strikers. These included the diesel and compressor operators, the blacksmiths, and the tool sharpeners. As a group, they held the highest-paying jobs in the mine. On the other hand, the tram and mill operators were loyal to a man. As the strike went into its fourth week, however, with no end in sight, they began seeking employment elsewhere.

Those on strike were ordered to vacate the company houses they were occupying, and the electric meters were removed from each house as it was vacated. Most moved to tents on Rivera's property.

A Mr. Matthewson, a Federal mediator, arrived at the mine early on the morning of February 20. He spent the morning with the union leaders, and the afternoon with Matson and Attorney Wilson. He left early in the evening after saying that he could see no room for conciliation. During the following week, the State Labor Commissioner attempted to work out a compromise, but he too left with the conflict no nearer to solution.

On February 28, a Mr. Brown from the headquarters of the union's international office in Salt Lake City arrived, and another meeting was called; the State Labor Commissioner was invited to attend. The position of the company was fully explained to Mr. Brown and he was told that as long as Gennis, Palleros, and the Ligue Obrera were involved, there was no point in continuing with further discussions. Brown left saying that the International would continue support of the strike.

On March 9, and again on March 12, Matson and Wilson met with Governor Tingley who was deeply concerned with the severe economic impact the strike was having on the state. When told that Palleros was one of the leaders, the Governor said that the situation looked quite hopeless. He said that it was Palleros who had led a group of relief workers in an invasion of the capitol the previous year.

On March 21, a Mr. Sherman called Matson and introduced himself as the Labor Commissioner from the Department of Labor. He requested a meeting, and after consultation with Wilson, one was arranged for March 23 in Santa Fe's Federal Building. Both Matson and Wilson were appalled that Sherman, who should have been neutral, was so obviously pro-union in the discussions that day and in the future. He began by insisting that no discussions should take place without Brown being at hand. The meeting was adjourned and another was arranged for the following day. In the morning meeting, Sherman would not proceed unless representatives from the union were present. Again, the meeting was adjourned until the afternoon when a committee from the union was there. It began at two p.m. and the following points were outlined by Wilson as a basis for further discussion:

1. Amco had no objection to recognizing the union but adamantly opposed a closed shop contract.

2. The union should concern itself strictly with legitimate union activities and not espouse communistic dogma to its members.

3. That union men would be welcome to serve on a grievance committee that had been functioning long before the strike began, and that several of those now on strike had in fact been serving on that committee.

4. That the company was totally opposed to making any agreement signed with the union binding upon all employees whether they did or did not belong to the union.

5. That the company would not deal with the union on any matter so long as Gennis, Palleros and the Ligue Obrera were involved.

At that point in the discussions, the meeting was interrupted by a man claiming to be the Secretary-General of Ligue Obrera. It was reported that, "in a loud and boisterous" voice, he demanded that he must be involved in any meeting in which discussions concerning the strike were being conducted. Finally, Sherman had some guards in the building eject the intruder, and he then asked the union people if they were in fact operating under the leadership of Ligue Obrera. Although they denied it, it seemed obvious that they were. It was hoped by Matson and Wilson that Sherman's attitude would change somewhat, since he had had a first-hand demonstration of what they had been dealing with. Unfortunately, they reported, it was not to be—in future meetings he continued in his pro-union stance even more belligerently.

The meeting was adjourned, with no progress being made toward resolving the situation. On Sunday, April 12, Dr. Elliott, Regional Director of the National Labor Relations Board called Wilson at home, asking to meet with him. Wilson invited him and the Board's attorney, Mr. Miller, to come to his house for the meeting. They met for about three hours discussing the whole conflict up to that date. Dr. Elliott requested a meeting the next day with Sherman and Matson present. After much discussion, Dr. Elliott had a paper typed which was presented as a basis for further discussion with the union. That paper read as follows:

IN THE MATTER OF THE AMERICAN METAL COMPANY AND
THE UNION OF THE METAL MINERS

If the union agrees to call off its strike and forego its demand for a closed shop and cooperates with the company toward improvement in efficiency, which is the basis for progress, both for the company and the workers—

1. The company agrees to open its operations as quickly as it can; re-employing such men as needed in these operations with regard to efficiency and without discrimination because of union or non-union activity.

2. The company will bargain collectively with the representatives of the union relative to working conditions, and it is the desire of the company to improve conditions as rapidly as production and proven resources of the mine warrants.

3. It must be understood that the agreement to bargain with the union is dependent upon the fact that the leadership must be responsible and reliable so that when an agreement is made that it may be lived up to by both parties.

4. The matter of increment in wages is entirely dependent upon the development of higher metal content ore. This development is underway and unless, and until, such development discloses ore in sufficient value and of commercial value with a per ton net unit of recovery better than the present ore, no increase in wages can or will be made.

With some minor modifications, the company was willing to entertain proposals one and four, but proposals two and three were rejected out-of-hand and the following was offered in their stead:

> It is understood that the Tererro mines of the American Metal Company of New Mexico will be operated on an open shop basis. This does not mean that the management is opposed to unions as such nor yet that it is opposed to a union being organized and conducted at Tererro, but it does mean that there will be no discrimination between the employment of union or non-union men, and that the men employed by the American Metal Company of New Mexico shall be free to join or not to join the union as each one sees fit. The management will meet with a committee or committees designated by the union for the purpose of presenting grievances or conditions affecting any union men, but the company will also entertain any committee or committees from any other organization of men employed by the American Metal Company of New Mexico and will give them equal consideration and this same treatment will be extended to any individual or individuals desiring to present their grievances or difficulties.

The following day, Brown was asked to rejoin the discussions. He proposed that the company re-employ everybody on the picket line, regardless of their former records. He was told that all but about twenty could be re-employed, but that there was serious concern about re-hiring the twenty, due to their unlawful activities over the past few weeks. A discussion followed as to whether the twenty would be named at that time or left to be designated at the time of re-employment. Brown wanted them to be named immediately, Matson refused to be specifically pinned down without the men's records before him, and the meeting was adjourned.

On Thursday, three striking members joined the discussions. The meeting ended with the strikers still opposed to any settlement unless all strikers were re-hired. Wilson had business in Phoenix and did not participate in further meetings. Prior to leaving, he, Matson, and Bemis again met with Governor Tingley, who promised his full support with State Troopers if the company would reopen the mine to production. He was still concerned about the negative impact the strike was having on the economy of the State.

Matson, Bemis, Hoag, Dr. Elliott, and Sherman met again on Thursday afternoon, April 16. About three hundred men were now working on development in the mine and related tasks, and management felt that they should have some say in the final outcome of the negotiations since they had been loyal employees and defied the strikers. Dr. Elliott concurred with the company's position (that those men currently employed should have some say as to the outcome of the discussions), and went to Tererro to discuss this with as many of them as possible. The consensus of all concerned was that, since there were about twenty strikers who had had serious confrontations with many non-strikers, they should not be re-employed because of the potential confrontations and physical violence which might occur after operations resumed.

On Saturday, April 18, another meeting was held at the mine, at which time Dr. Elliott submitted the following for consideration:

AGREEMENT OF SETTLEMENT OF THE CONTROVERSY BETWEEN THE AMERICAN METAL COMPANY AND THE TERERRO MINERS UNION LOCAL 64

April 18, 1936

1. No medical examination will be required of the men re-employed who were formerly on strike.

2. There will be no charge for replacement of meters.

3. There will be no wage reductions to any of the men re-employed.

4. There will be no discrimination against men because of union activity; however the company reserves the right to hire and fire.

5. The men will be replaced in the houses formerly occupied by them in every instance where possible.

6. It is uinderstood that the Tererro mines of the American Metal Company of New Mexico will be operated on an open shop basis. This does not mean that management is opposed to unions as such nor yet that it is opposed to a union being organized and conducted at Tererro, but it does mean that there will be no discrimination between the employment of union or non-union men, and that the men employed by the American Metal Company of New Mexico shall be free to join or not to join the union as each one sees fit. The management will meet with a committee or committees designated by the union for the purpose of presenting grievances or conditions affecting any union men, but the company will also entertain any committee or committees from any other organization of men employed by the American Metal Company of New Mexico and will give them equal consideration and this same treatment will be extended to any individual or individuals desiring to present their grievances or difficulties.

7. The credit coupons now in use will be discontinued.

8. All men now on payrolls will be retained, subject to the company's employment regulations.

After much discussion, Brown and the strike committee left the meeting and agreed to return at 3 p.m. Brown returned by himself and said that they still rejected the settlement as proposed by Dr. Elliott, as well as the company stipulation that the twenty men would not be re-hired. He agreed to return again at three on Monday afternoon.

On Sunday, Sherman called Matson to say that the men currently employed should have no say in the outcome of the negotiations. If that was going to be the stance of the company, he considered that it was allowing the employed men to run the business and, as such, they were obstructing the Government from trying to make a settlement. This was, of course, not Dr. Elliott's opinion, since he had personally talked to the men employed to find out their feelings about the matter. It seemed to be a serious philosophic difference between the two branches of the Government.

Brown and the committee returned as promised on Monday. They seemed to be more willing to accept the terms in Dr. Elliott's conditions submitted on Saturday, but felt that the union was not mentioned enough. Meanwhile, Dr. Elliott had prepared yet another proposal for consideration as shown below:

STATEMENT CONCERNING POSSIBLE STRIKE SETTLEMENT
AT
TERERRO, NEW MEXICO

1. If the union is really interested in getting men on strike back to work, and also those who have left here for whatever reason, the strike should be called off, the picket lines removed, and thus give the company a fair chance to work out a plan of re-employment. In case the strike is called off, the majority of the men now in the picket line will be re-employed as promptly as possible, as well as all other former employees. The company reserves the right to hire and fire according to its employment regulations.

2. It is understood that the Tererro Mines of the American Metal Company of New Mexico will be operated on an open shop basis. In order to fully promote harmony between the company and any group or groups of employees, it is agreed that the committees from any group, or that any individual will be met with for the purpose of discussing grievances.

3. No medical examination will be required of the men re-employed who were formerly on strike.

4. There will be no charge for replacement of meters.

5. There will be no wage reductions for any of the men employed.

6. The men will be replaced in the houses formerly occupied by them in every instance possible.

7. The credit coupons now in use will be discontinued.

8. All men now on payrolls will be retained, subject to the company's employment regulations.

After considerable discussion, the committee decided to reject the whole thing because of the statement in item one that "...the majority of the men" would be re-employed. Since there were about 150 men now on strike, they took "majority" to mean that as many as 74 could be refused re-employment. In order to satisfy their objections, and to re-insert item 6 desired by the company, a final proposal was drawn up by Dr. Elliott as follows:

AGREEMENT OF SETTLEMENT OF THE CONTROVERSY THE AMERICAN METAL COMPANY AND THE TERERRO
MINERS UNION LOCAL NO. 64

APRIL ?, 1936

1. No medical examination will be required of the men re-employed who were formerly on strike.

2. There will be no charge for replacement of meters.

3. There will be no wage reductions for any of the men employed.

4. There will be no discrimination against men because of union activities, after this settlement, so long as these activities are not conducted during working hours or while on duty. The company reserves the right to hire or fire.

5. The men will be replaced in the houses formerly occupied by them in every instance possible.

6. It is understood that the Tererro mines of the American Metal Company of New Mexico will be operated on an open shop basis. This does not mean that the management is opposed to unions as such nor yet that it is opposed to a union being organized and conducted at Tererro, but it does mean that there will be no discrimination between the employment of union or non-union men, and that the men employed by the American Metal Company of New Mexico shall be free to join or not to join the union as each one sees fit. The management will meet with a committee or committees designated by the union for the purpose of presenting grievances or conditions affecting any union men, but the company will also entertain any committee or committees from any other organization of men employed by the American Metal Company of New Mexico and will give them equal consideration and this same treatment will be extended to any individual or individuals desiring to present their grievances or difficulties.

7. The credit coupons now in use will be discontinued.

8. All men now on payrolls will be retained, subject to the company's employment regulations.

The committee agreed to put the agreement to a vote of the entire union and report back the following morning. Accordingly, on Tuesday morning, April 21, the committee reported that the union had rejected the whole matter unanimously. Dr. Elliott decided the dispute was not arbitrable and returned to his headquarters in Fort Worth. Sherman had already departed and negotiations were, once and for all, ended.

The following morning, the company mailed notices to all those who had left the area, and for whom they had forwarding addresses, to advise them the strike had been put down, and to ask them to return to work. Jobs were also offered to those on the picket line. Gradually, and grudgingly, men dropped out of the line and asked to be re-employed.

Hoisting of ore and operation of the crusher began on Thursday morning, April 23—ten weeks and two days after they were shut down by the strike. No concessions were won by the union, nor were they recognized as a bargaining unit for Amco's employees. Almost all employees, strikers and non-strikers alike, lost 20 percent of their annual income. The economy of the State was seriously affected and the future of Amco's mining operations was seriously jeopardized. Two basic changes resulted from the strike—the mining complex was placed off-limits to all those who were not reporting for, returning from, or at work, and the issuance of coupons was discontinued.[2]

By May 15, 1936, the mine was back in full operation with a full complement of employees. How long that operation would continue was fully dependent upon whether or not more high-grade ore could be located. The known remaining ore reserve at that time was about 775,000 tons, enough for no more than four more years of operation. Continued operation during those four years would be tenuous at best. Any further decrease in price of lead or zinc, any increase in the cost of labor, or any other interruption in production would be the death knell signaling the end of the mining operation. That event occurred on October 1, 1938.

---

2. Picks' General Store operated, at Amco's request, on a semi-monthly credit basis. Employees who charged purchases (the majority did) were expected to pay all charges by the 10th and 25th of each month, five days after payday. Many did not and would get hopelessly in debt. To overcome this, the company began issuing coupons in lieu of pay checks which could only be cashed at Picks', thereby forcing them to pay their charges before using the funds for something else. If the employee had more than enough to pay for charges at the store, they could exchange the remainder for cash. However, not all employees were paid in coupons—some received pay checks depending upon the importance of their job.

chapter 14

* *

May 31, 1939—the Venture Ends

On March 1, 1937, Matson reported to the New York office that the remaining ore reserve was about 260,000 tons, and worthy of continued mining. Prices for lead and zinc had risen somewhat from the previous year, therefore further exploratory work would be pursued. One year later the reserve had increased by about 4,600 tons; 186,425 tons had been mined during the ensuing year.

However, the reserve below the 1,200 level was only 43,000 tons. Exploration during the year had not led to the discovery of additional reserve in those lower levels. Only 3,289 tons had been mined from the lower levels, at a cost of $48.19 per ton, as compared to $7.71 per ton for that mined above the 1,200 level. It seemed prudent, therefore, to cease operations below 1,200 and concentrate all effort on the levels above.

All previously blasted ore was removed (5,115 tons) by mid-April, 1938, and removal of all equipment in the lower levels was begun. All had been removed except the pumps by June 12; they were shut down at eight a.m. My father was part of the crew selected to remove them. The crew's instructions were to attempt to get the pumps removed before the water level rose to their chests—if they could not, the pumps were to be abandoned in place. The water had risen to their waists when the hoist operator was signaled to lift them out at 12:15 p.m. By June 15, the lower levels were completely flooded.

In June of 1937, the Federal Wage and Hour Law, sometimes referred to as the Wagner Labor Act, had been introduced in Congress to establish minimum wages and maximum hours for employees engaged in work associated with interstate commerce. If the law passed, Matson wrote to the American Mining Congress in Washington, a lobby for the mining industry, it would be the death knell for the Pecos Mines. At that time there were 857 full-time employees. See Appendix D Miners earned an average of $4.25 per eight-hour day, muckers $3.55, and laborers working on the surface, $3.00. Most worked an average of 52 hours per week, with a few who might work 56 hours.

In his letter, Matson pointed out that a change to a 40-hour week, with time-and-one-half for hours worked between 40 and 48, and double-time for hours worked over 48, would be impossible to meet given the circumstances. To continue to operate at the same level of production by working men only 40 hours per week with no overtime, would require an additional 257 employees. Not only would this be unachievable, because experienced miners were almost impossible to find, the company could not afford to build additional housing required for those new employees and their families.

On the other hand, if no new employees were hired, and the existing employees were paid the required time-and-one-half and double-time for their average 52-hour weeks, the cost of labor would rise by 15.4 percent, a prohibitively high increase, since the net result would be an operating loss—one that the company was not willing to take.

Congress passed and President Roosevelt signed the bill in September, 1937; it was to become effective on October 1, 1938. One provision of the law made it illegal to hire anyone under 18 years of age in a hazardous industry, mining being one of them.[1] The overtime provisions of the law would affect all employees eight months later—because of the increased costs, a decision would be made to close the operation, and all but a handful of the employees would be laid off. Even that handful would work only until the mine and mill were totally shut down and all assets disposed of.

All further development work in the mine was stopped in June, 1938, except in ore bodies already being worked, or to open known adjacent ore bodies. All diamond drilling had been stopped the previous December. Because mining had ceased on levels below 1,200 in June, the cost of mining had dropped an average $1.24 per ton for the year 1938, but labor costs had increased by 3.85 percent per average man-hour worked during the year, thereby partially negating the savings on mining costs. Any hope of realizing a profit in 1939 had vanished, and a decision was made in early March to close the operation.

Notice to employees of the pending cessation of operations was withheld until May 20; on midnight, May 31, all operations were shut down. A few employees were retained to remove equipment from the mine, but it was difficult to get even those few to stay behind. Most were anxious to beat the other fellow to potential jobs at other mines, principally the potash mines in Carlsbad. The venture had lasted five years and five months longer than had been expected when operations began twelve years, five months earlier.

The operation had mined 2,293,582 tons of ore, had milled 2,299,082 tons[2] which refined to 178,200 ounces of gold, 5,642,627 ounces of silver, 19,296,691 pounds of copper, 138,411,920 pounds of lead, and 440,683,250 pounds of zinc.

At the mine, about six and three-fourths billion gallons of water had been pumped to the surface, while at the mill, almost three billion gallons of water had been pumped from the Pecos river to provide for domestic use, ore processing, and power generation. About 180 million kWh of power had been produced.

Although no figures could be located which dilineated the exact costs associated with the preparation of the mine and mill for production, an analysis of various documents concerning cost of labor, mandays worked, contract costs, direct purchases, general correspondence, media reports, and etc. indicated that almost $3,150,000 was

---

[1] I would be the only person at the mine affected by that provision of the law since I was the only employee under 18 when the law went into effect.

[2] The difference of 5,500 tons is accounted for by the tons of ore milled that had been stockpiled prior to start-up of operations. See Appendix A

expended before the first ton of ore was mined, milled, and shipped to a smelter. An approximate breakdown of those costs follows:

51% interest in the mine	$1,250,000
Aerial Tramline	330,200
Milling Plant	495,000
Power Plant	265,000
Pumping Plant	54,000
Mine Industrial Plant	352,000
Mine Development	266,000
Mine Camp Construction	68,000
Mill Camp Construction	23,000
Road Construction and Improvements	11,000
Rail Spur	13,600
Miscellaneous	22,200
Total Estimated Cost	$3,150,000

Any single bucket on the tramline would have transported 5734 tons of ore and made 9174 round trips from the mine to the mill. At 23.48 miles per round trip, each traveled almost 215,500 miles, only about 22,500 miles short of a trip to the moon. All 400 buckets traveled a total of 86,165,288 miles, just 7,000,000 miles short of a trip to the sun. See Appendix A and Bid Parameters, p. 20

Over the life of the mine, an annual average of 600 men and their families had been provided employment, especially during the worst depression in our nation's history. For each ton of ore mined, 1.29 mandays of work was expended. See Appendixes A & D At an average daily wage of $4.00, the total cost of labor was $11,833,752, and the average earned by anybody working the full 12 years, 5 months was $19,723, or a total of $1588 per year. The start of WWII in Europe, three months later, would signal the end of the depression, jobs would become plentiful, and the standard of living for everybody would begin to improve.

chapter 15

Gold Fever—the Search Goes On

Operations ceased on Wednesday, and on the following Monday, dismantling of the equipment and all structures at both the mine and the mill began; except for four of the original five ranch houses, everything would be gone by February, 1942. Since the mine property was located within the Pecos River Forest Reserve, Amco had made an agreement with the U. S. Forest Service that, upon cessation of operations, the area would be returned to its natural state insofar as possible. Because of the agreement, everything had to go.

There were only a handful of employees retained on the payroll to accomplish the work—none were married. Most of the dismantling work had been contracted out, and the majority of the structures were sold to a salvage company that immediately began tearing down the vacant houses, both company and privately owned. The Pecos road had never been so busy, what with a steady stream of trucks arriving empty daily and driving away piled high with salvageable lumber or with tarpaper and other junk to be disposed of somewhere outside the National Forest. Three people remained in the office to handle the business of closing down—Matson, Russell, and Les McClure.

Les McClure's parents and his sister and brother-in-law from Dallas were contracted to operate a restaurant in the old messhall so as to provide a place for the remaining employees and contract workers to eat. Joe Matson, Jr. and I were given a contract to operate a service station at the site of what had been the company's vehicle maintenance shops across the road from the restaurant. This was done in order to make gasoline available for the company's few vehicles and for the remaining employees' automobiles. However, we also had the option of selling to the summer residents of Holy Ghost Canyon, Cowles, and other cabins along the upper Pecos, and to itinerant motorists passing through the abandoned town.

On August 15, Mrs. Matson and Alice moved back to Denver; Joe left on a trip with his father to Shafter, Texas. Matson returned only to arrange to have their belongings moved to Santa Fe where he went to set up an office. Russell remained in charge at the mine, since he and his wife lived in a home they had built in Holy Ghost Canyon.

On August 24, I closed down the service station and closed out my accounts with Mr. Russell. The following day, he drove me to Glorieta to catch Santa Fe's *El Capitan* to California. On September 1, the restaurant closed and Les's family moved back to Dallas. For all intents and purposes, except to finish dismantling equipment and disposing of it, the mine ceased to be a major employer, and Tererro became a ghost town.

On April 1, 1940, the payroll was reduced to four employees—Matson and Russell in an office on Sena Plaza in Santa Fe, Bill Orton at the mill, and Roy Roberts at the mine. Orton and Roberts acted as watchmen and showed prospective customers the salvaged equipment. They received 45¢ per hour and worked a seven-day week. The total cost for their labor was about $245.00 per month. Total monthly expenses amounted to about $745 per month, a figure that included a part of Matson's salary, all of Russell's salary, the two watchmen's wages, office rent and its expenses in Santa Fe. This expense would continue until all equipment had been sold or disposed of.

By November 15, 1940 the value of the equipment yet to be sold was estimated to be $103,965, $25,000 of which was for the tramline system, $75,000 for all other equipment, and $3,965 for new and unused mine and mill warehouse supplies. Norcross was in the process of getting a new mining operation underway in Cuba, and had requested that the tramline not be sold until it was determined if it could be used there.

An inventoryy taken on October 15, 1941, revealed that there remained at the mine 24 buildings, including the hospital, Matson's residence and associated buildings, the warehouse, the blacksmith shop, Picks' store and 12 other residences. The crushing plant, the power line along the tramline right-of-way, and the tramline itself remained in place. By February, 1942, all of the buildings except Matson's residence, the hospital, and two residences on Davis Creek, had been sold and removed for a total of $4,050. The tramline, power line, and crushing plant had been dismantled and moved to the mill where they would be easily accessible by rail for anyone who wished to purchase any of it. The majority of the mill equipment had been sold to the Cotopaxi Exploration Company and was being shipped to Guayaquil, Ecuador. Estimated value of the remaining equipment from both the mine and mill was set at $30,000. Records of when and to whom this remaining residue from the mining/milling operation was finally disposed of could not be located.

Matson received a report from the New Mexico State Engineer on November 12, 1941, of his department's inspection of the two tailing dams at the mill. The report concluded by stating, "Failure (of the dams) will cause no dangerous condition downstream because of the water or its impact. It is possible that a silt problem on downstream lands might be created. It is recommended that these dams be stricken from our inspection itinerary."

Upon cessation of production, all of the mine properties were conveyed by Amco to Pecos Estates, Inc., a New Mexico Corporation. On June 6, 1950, Pecos Estates, Inc. conveyed all of the mineral rights in those lands to the Pecos Mineral Trusteeship under a Declaration of Trust establishing three trustees, Mr. John Payne Jr., Mr. Thomas Moore, and Mr. Fred Norcross Jr., thereby leaving Pecos Estates holding only surface title to the land. The trust consisted of 300 shares, 153 being owned by American Metal Climax Corporation, successor to American Metal Company Ltd., and known today as AMAX Corporation; the other 147 shares were owned by the group headed by David Goodrich, this being in the same 51/49 ratio as had been the prior ownership of Amco. Norcross represented the Goodrich group, and the other two trustees represented American Metal Climax. Amco had been dissolved as a corporation on December 13, 1945.

At the time the trusteeship was established, the Game Commission of New Mexico purchased all of the shares of Pecos Estates, Inc., then dissolved Pecos Estates, thereby leaving the Game Commission as owner of surface title. In 1962, the Forest Service negotiated an exchange of land with the Game Commission, in which the Forest Service acquired what had previously been the Marlow and Chapin tracts of land.

Upon transfer of the mineral rights to the three trustees, Matson was instructed by the trustees to make sure that there was no possibility for anyone ever to be placed in danger as a result of those mineral rights. The old Evangeline adit was still open, as was a smaller adit, somewhat higher up the mountainside, that had been opened to explore the contact between the pre-Cambrian and overlying sediments. Matson hired one of the miners still living in Pecos, and the two of them spent two days blasting and caving the lower adit until it was totally closed.[1] The main shaft had been closed at its collar with a 8-inch-thick bulk-head of concrete in 1940 and had shown no signs of deterioration.[2] The openings into the mine from the two glory holes had also been blasted full of rock in 1940, so the old mine workings were deemed to be totally secure and free from potential liability from invasion by any person.

On June 30, 1972, the trustees (Erwin A. Weil had succeeded Fred Norcross, Jr.) of the Pecos Mineral Trusteeship granted to a Nevada corporation, Perry, Knox, Kaufman Inc., a five year option to purchase those mineral rights and the sole right to conduct mineral exploration upon the lands held in the trust. In exchange for the option and the right of exploration, the trust was to receive 15 percent of the net profits resulting from any further exploitation of the holdings. The optionee was further required to expend not less than $30,000 by the end of the second option year, $80,000 by the end of the fourth option year, and $300,000 by the end of the fifth option year in further exploration work. Failure to achieve any one of the three expenditure requirements was to result in default of the option 45 days after the required date of compliance. A six month and a three month extension was granted to the optionee by the trustees when they, the optionee, failed to meet the first $30,000 goal. They did expend the $30,000 by end of the second extension, but the option was dropped prior to the expiration of the fourth option year.

Beginning in 1975, when the American public was again permitted to own gold and its value began to climb, there was further renewed interest in the Pecos Mines and the possibility that there might be an advantage to exploring the area further for additional ore bodies. In 1977, Mr. W. D. Riesmeyer, an Albuquerque geologist seeking a Master of Science degree at the University of New Mexico, entitled his thesis *Precambrian Geology and Ore Deposits of the Pecos Mining District, San Miguel and Santa Fe Counties, New Mexico.* In it he stated that there was a distinct possibility of undiscovered ore bodies below the 1,700 level at the abandoned Pecos Mines. He further suggested that the presence of such ore bodies could easily be determined with a modest drilling program from the surface.

The Continental Oil Company (Conoco) had launched a drilling program known as the Pecos Project in the area in 1977. A report of that program was submitted to its Metallic Division's annual meeting at Bishop's Lodge in Santa Fe on April 30, 1978.

---

1 Although records indicate that the upper adit was also to be blasted until inaccessible, as of this writing, it is closed in only with timbers and a heavy wooden door.

2 Although records show that the bulkhead was to have been poured five-foot thick, examination in later years show it to be only eight inches thick, but that there are several other similar bulkheads down the shaft.

Essentially, the drilling program was concentrated on the Jones Mine lease in Macho Canyon.[3] It was concluded in the Conoco report that both the Jones Mine and the Pecos Mine deposits lay in the same "horizon," thereby opening many exploration possibilities in the area between the two mines.

Another mining engineer, Nelson King, had issued a report dated September, 1979, concerning the possibility of further exploration of the Christino Rivera Mining Company's[4] mine located on the Cowles road adjacent to what had previously been Amco's property. The Company was formed when it became known that Amco was going to open its mine in 1926. A Mr. Paul Gullot was President of the corporation for which he had provided the financial backing; Rivera was a partner having provided the surface and mineral rights.

The Company opened and attempted to develop the mine during the same time the Amco operation was underway, and at one time attempted to sell the claim to Amco. A shaft was sunk to 537 feet which passed through veins of ore at the 100 and 300 foot levels. No further ore was encountered in the remaining 237 feet. No development work was carried out in the ore bodies, since Gullot was determined to reach the 500-foot level before starting development. He charged at one time that Amco encroached upon his claim, a charge that was denied by Amco.[5] Gullot died unexpectedly shortly after Amco closed the Pecos Mine. Because of his untimely death, and the fact that their shaft began filling with water when the Pecos Mine was allowed to flood, the company brought all mining and development activity to a halt, and it has never been resumed.

In King's report, he cited both the Conoco report of April 30, 1978, and the Riesmeyer thesis as evidence that someone should reopen the Rivera mine because of its considerable precious metal potential. The following is excerpted from his report:

> The Pecos Mine survived through the period of the Depression and produced nearly 2.3 million dry tons of massive sulfide (lead-zinc) ore with minor amounts of gold and silver from 1927 to May 31, 1939. The abrupt closure of the facilities was attributed to depressed metal prices, decreasing grade, high-cost of production (square-set), water problems and occasionally violent labor problems. It is interesting to note that production was at a maximum in 1938 with early 1939 level of production not far from the average of previous years. The abrupt closure of the Pecos Mines remains a mystery to this writer because of the generally improved economic outlook in 1939 and little indication of

---

[3] John Jones, an Amco employee, and a friend of my father, discovered an outcropping in Macho Canyon and filed claim to it. He began mining the claim on a small scale during his spare time, but removed only a minimal amount of ore. He eventually sold his interest to the St. Cloud Mining Company.

[4] Christino Rivera was the father of Louis Rivera who owned the property on the valley floor.

[5] Matson's background was not as a mining engineer, but as a surveyor. As a young man, he had made a name for himself when he worked for a mining company in Tonapah, Nevada. He had been instrumental in settling a similar disagreement between two other mining companies by skillfully surveying both properties above and below ground, and proving that the one company's drifts stopped just inches short of the other property. Based on his surveying skill, American Metal hired him at their Climax operation, and he quickly rose to general manager of the operation. Because of this prior experience, it is doubtful that he would have permitted any encroachment into the Rivera property.

decreasing production or other associated problems on a scale sufficiently detrimental to force the closing of the mine.

Perhaps the information provided in this book will clear up the mystery for King. At best, the mine could have remained in production another year, but it would most certainly have operated at a loss during that year. Although King states that the production was at maximum in 1938, the 1929 production exceeded it by about six and one-half percent, and the 1939 production was about five percent below the previous 12 year average. See Appendix A What King does not cite are the net losses sustained during those years—$15,546.00 in 1937 and $73,327.00 in 1938. I could find no figure for 1939, but it is almost certain that those net losses were continued into the last five months of operation. This was reason enough to shut the operation down even without the large cost increase resulting from the new Federal Wage and Hour Law.

On February 13, 1979, the Continental Oil Company (Conoco) obtained a 30-year lease from the trustees in return for which they were to expend the following on exploration, development and mining operations: $10,000 the first year, $25,000 the second year, $50,000 each of the third and fourth years, $65,000 the fifth year and $100,000 each of the next five years. In return for the lease, Conoco was to pay the trust either six percent of the sales price of the mined ore if it sold for less than $50 per ton, or eight percent of the sales price if it sold for more than $50 per ton.

On December 7, 1982, Conoco sold both leases for $10.00 to Santa Fe Mining, Inc. (later changed to Santa Fe Pacific Mining, Inc.), a Kansas corporation. With respect to the Pecos Minerals Trusteeship lease, they had expended $178,236 of the required $200,000 by February 13, 1984. Therefore, Santa Fe Mining was required to spend an additional $21,764 before that date. Riesmeyer, who was then with Santa Fe Pacific Mining, Inc., again issued a report in August of 1987 in which he concluded that there had to be potential gold mineralization at rather shallow levels below the outcroppings that Case had discovered over 100 years earlier, also he felt that there might be a deeper potential in the hanging wall of the Katy Did ore body to the southwest. He recommended, "....the Pecos Mine (AMAX lease) be farmed out for further drilling and exploration." Although he used the term "AMAX lease," he should have referred to it as the Conoco mineral rights lease (as described earlier) from the Pecos Mineral Trusteeship.

Santa Fe Pacific Mining did in fact meet all requirements of the lease until May 2, 1990, when they issued and had recorded in the County of San Bernarlillo, a termination of the lease During the term of the lease, they had entered into a joint venture agreement with Noranda Exploration Company from Denver to carry out exploration on the Jones property (then owned by St. Cloud Mining); on the Pecos Minerals Trusteeship property; and on the Christino Rivera property just north of the Tererro mine. Noranda issued a report of their work in June of 1989, and recommended that all leases be terminated due to insufficient mineralization to justify further expenditures.

It seems reasonable to believe, therefore, that the final chapter of the Tererro mining venture seems now to have been written. Since every effort to find sufficient ore reserves in the vicinity of the original mine failed to justify a new mining venture, it would appear that any further financial investment seeking the illusive gold would be both foolish and impractical unless some radical new methods of mining and milling are someday developed..

THE GLORIETA DEPOT
Highest Station
On Santa Fe's System

Above

WOODEN TRUSS BRIDGE NEAR HOLY GHOST CREEK
The first structure built by AMCO
and the only structure still standing.
Building at right-center is Tererro General Store
and the modern day
Tererro Post Office

Left

MEADOW BETWEEN TWIN BRIDGES,
(OUR FIRST ISOLATED TENT-HOME SITE)
Nothing remains of the bridges and the pine trees now cover the meadow.

North
MINE COMPLEX
(Looking)
East
South

Ole Lee, Master Mechanic
H. L. Brown, Denver Office Manager
J. T. Matson, General Manager
C. Hoag, Mine Superintendent

H. D Bemis, Mill Superintendent
J. T. Matson, General Manager
Charles Stott, Chief Mining Engineer, Shafter Mine
Mrs. Stott

Harold Walter
Ray Marsten
Checking Tramline
After Heavy Snowfall

Milling Complex at Alamitos

Tramline
Looking North from Mill

Tramline Repair Crew

Tramline Tension Station
South Ridge Above Dalton Canyon

The DUMP, the DEERLICK, the TERERRO, or the FROGPOND

Facing page—aerial view of ALAMITOS and MILLING PLANT

The INDIAN CAVE

TERERRO
Fisheye View

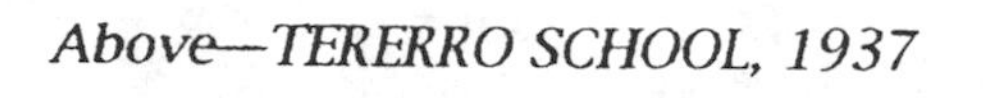

Above—TERERRO SCHOOL, 1937

Right—OUR TWO ROOM HOUSE, 1926

Below—MATSON HOUSE, 1933

Left

ALICE MATSON BIRTHDAY PARTY
Left to right-Betsy Anderson, Lorraine Smith, Alice Matson, Joe Matson , Author, Dick Smith, LaVoyse McDuff

Right

JOE MATSON BIRTHDAY PARTY
Front row:
Herb Bemis, Joe, Fred Bemis
Author, Dan Smith
Back Row:
Delta McDuff, George Smith, Unknown, Dick Smith, Richard Wheelock, Norman Littrell

Left, part of

SECOND, THIRD, AND FOURTH GRADES
1927

Author and Joe, front row, right
Jane McDonald, front row, left

Right—SPECTATORS

Delta McDuff, Author, Norman Littrell Mrs. Hoag, Miss Parsons

Below—TEEING OFF, # 6

Jack Finlay and Dr. Smith

Above—THE LONG DRIVE

Delta McDuff on #3 Tee

Left— A FEARSOME FOURSOME

Richard Wheelock, Sr., Ole Lee, Eugene Anderson, Unknown

Above—SIMMONS RANCH
before being converted
to hospital, 1925

Left—AUTHOR
leaving hospital after having
quarter implanted in his head

Above, clockwise—

AUTHOR AND JOE MATSON , 1927

PAT MONTOYA, AUTHOR and JOE, 1932
at clubhouse built from purloined materials

LEAVING ON FIVE WEEK WILDERNESS TRIP, 1937

PROUD GAS STATION PROPRIETORS, 1939
restaurant is in background

Left—AUTHOR, 1988
plaque is in memory of Joe whose ashes were scattered beneath tree

Above—PACKING IN TO PECOS WILDERNESS
Below—MAY WALTER, BELOVED TEACHER, FRIEND

The McDuff Family, September 1926
Picture was taken standing by Pecos River
at site of second tent home.

DOWNTOWN TERERRO, 1930

part 2

✲ ✲

The Town

✲ *its folks*

✲ ✲ ✲ *its features*

✲ ✲ ✲ ✲ ✲ *its faults*

chapters
16 thru 25

chapter 16

* *

An Isolated Meadow Home

My father paced up and down the platform at Santa Fe Railroad's Glorieta, New Mexico, depot, the highest town on its system. He anxiously awaited our arrival; we were no less anxious, for it had been over five months since we had last seen him. On this day, May 8, 1926, my mother, my two sisters, my brother, and I[1] were moving to a new home—a nameless place on the upper Pecos River.

Dad had traveled to this nameless place in late December of the previous year to seek employment with the American Metal Company of New Mexico that had just announced they were going to open a mine there. He was lucky; he was hired as Construction Supervisor—the third person to be hired by this new company. His first assignment had been to employ a crew and oversee the construction of four heavy-timber, truss bridges across the Pecos River. The cold winter/early spring months were coming to an end, and he decided he wanted his family with him at the mine-site, although no houses had yet been constructed.

We left Mountainair, New Mexico, our parent's home since 1915, early on Saturday morning, May 8. We travelled by train, and after a three-hour layover in Albuquerque, our scheduled arrival in Glorieta was around two p.m. Billy Montgomery, Supply Supervisor at the mine, drove Dad to meet us. He drove the company's Model "T" truck down the dusty, graveled, corrugated road to Pecos, and on to Glorieta on the more modern, paved highway US85. While he paced, little did Dad realize that we were on the train only by the good graces of the Albuquerque Police Department.

During our layover, Mom decided she wanted to go shopping in downtown Albuquerque, just a few short blocks from the train station. Not wanting four children along delaying things, she left us at the station waiting room with Balpha, age nine, in charge. She admonished us not to leave the waiting room for any reason. She told us three younger ones to mind our sister, otherwise we would suffer the consequences of a good spanking.

When she had not returned by the time promised, Balpha became concerned and decided we should all go look for her. Delta, age seven, refused; he remembered the

---

1 All of us, except me, had rather unusual given names. Dad's was Coyel Ferdinand, Mom's was Hartie Ann. My oldest sister's name was Balpha Adelia; my older brother's, Delta Otis; and my younger sister's, Lou Ella LaVoyse. Mine was more common, Leon Eugene.

admonishment not to leave for any reason. LaVoyse, age three, and I, age five, went with her; we remembered the admonishment to mind Balpha.

Just a few minutes after we left on our fruitless search, Mom returned. Delta explained what had happened; Mom called the police. They immediately launched a search, and found us aimlessly wandering in and out of stores. Our train had already pulled into the station. The conductor was calling, "All aboard", when the police squad car, red lights flashing and siren wailing, delivered us back to the station. We had to board immediately, which saved Balpha from her promised spanking for leading our misadventure into the big city.

We arrived in Glorieta on schedule. After a joyful reunion with Dad, the men loaded the baggage, and tucked the children into the bed of the truck. With the three adults in the cab, we headed for our new home, generally about a one hour drive in those early days. The distance was only 20 miles. Except for the frightening ride in the police squad car, this was the first-ever ride in an automobile for us children. The cool, clear, invigorating mountain air made it an exhilarating trip.

About three miles north of Pecos, a large boulder had rolled onto the roadway, blocking traffic in both directions. Try as they might, the men were unable to move it without some kind of help. After a search of the hillsides, they found a couple of small logs with which they could leverage the boulder to the edge of the road. After a delay of about two hours, they finally sent it crashing down the mountainside, flattening small trees and shrubs in its relentless journey to the water below. It met the river with a huge splash. That event planted a seed in my mind which would get me into trouble about three years later.

As we drove up the canyon, the mountains became higher, the canyon walls steeper, and the new spring vegetation more lush. Occasionally, we caught glimpses of Pecos Baldy, perhaps 35 miles away. In the clear, brisk air, it seemed close enough to touch. To us, the mountain scenery was spectacular; our previous scenery had been the flat bean fields around Mountainair.

En route we crossed the four new bridges built by Dad's crew. Just beyond the first, we crossed Dalton Creek, named for the infamous Dalton brothers who had a hide-out in the canyon in the early 1890's. Next were Macho Creek and Macho Church, built from native fieldstone by Franciscan friars around 1875. They were there to minister to the few Mexican families who had scratched out small fields in the relatively few tillable areas along the river banks.

Between the second and third bridges, we inched by Cathedral Rock, a huge outcropping of tan granite rising almost vertically some 500 feet above the roadway. There was only room for one-way traffic between the rock and the river, and it would remain that way until widened by the Civilian Conservation Corps in 1933-34.

Just beyond the third bridge, we crossed Indian Creek, where the valley floor widened into a half mile long, relatively flat meadow-like area dotted with stately ponderosa pines. At the southern end was Irvin Ranch—one of the first dude ranches in the area. At the northern end were several log cabins—most were built, owned, and occupied only in summertime, by oil-rich folks from Texas and Oklahoma, but one was owned by Charles Ilfield, who operated a lumber supply store in Las Vegas. There was one small pond at Irvin Ranch, and three more at the cabins on the north. These would provide winter fun for many hours during the next 13 years.

The fourth new bridge (the first to be built) was at the confluence of the Pecos River and Holy Ghost Creek. The southern boundary of the land owned by Amco traversed the valley floor just below this bridge. About 250 yards north of the bridge, there was a large limestone cave in the mountainside. Located on the west side of the river, this cave, we would learn later, played an important part in the rituals of an Indian tribe that had once occupied an abandoned pueblo about two miles south of the village of Pecos. Delta would become involved in one of those rituals some years later.

Between the cave and the river, Littrell's Dairy would very soon begin operation. The roadway, at this point, made a sweeping right turn away from the river. It went up the small Davis Creek Canyon for several hundred yards, made a "U" turn across Davis creek, and switched back and forth up the mountainside. It leveled off about 400 feet above and an average quarter-mile east of the river.

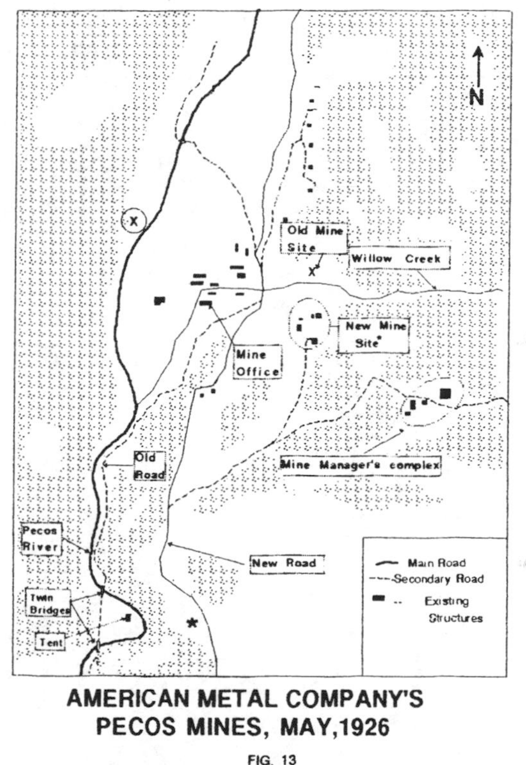

AMERICAN METAL COMPANY'S PECOS MINES, MAY,1926

FIG. 13

Originally, the road had continued straight ahead from the cave area, parallel and adjacent to the east side of the river. Our journey took us up this old, abandoned roadway. About a mile north of the cave, a portion of the mountain on the west side jutted out, and the river made a sweeping horseshoe bend around this protrusion.See Fig. 13 On the outside of the bend, the river washed against a precipitous mountainside—there was no room for the road.

Within the horseshoe, however, there was a small, slightly inclined meadow, about two acres in size. The road bridged the river on the south side of the meadow, traversed it on its west side, and bridged the river again on the north side. The Davis Creek route had been built to avoid the necessity of replacing these two log bridges. Also, it brought the road up to the approximate level of the new mine-site being developed.

This meadow was our final destination, for Dad had selected it as the site of our new home. It was accessible only by crossing over one of the two old log bridges known as the twin bridges. We approached the meadow on the old road from the south; it would be the last time any automobile would ever cross that south side bridge.

Dad had pitched a 10-foot by 14-foot tent, with a floor of pine boards. In one front corner was a small wood-burning cook stove, in the other a table and four chairs. In the rear corners were two beds, one for our parents and the other for all four children.

Outside, one path led to the river about 100 feet away, another led to an outhouse about 100 feet in the opposite direction.

We arrived around five p.m. It was not long before we settled in. Although darkness was still a couple of hours away, the meadow had been in the shadow of the mountain for over two hours. Dad built a fire in the stove, and Mom cooked our first meal in our new home—fresh mountain trout which Dad had caught before leaving for the depot earlier that morning. Darkness soon enfolded our little tent home, one where we would live for the next four months. A gasoline-fueled, silk-mantled lamp was our source of light; there would be no electricity at the mine-site until December 14, when the first electric power would be delivered from the new power plant at the mill.[2] Power for home lighting would not be available until the night of December 23. We all went to bed after having had a busy and exciting day. Exploring our new meadow surroundings would be high on tomorrow's agenda.

---

[2] A facility to mill the mine ore was being constructed simultaneously with the mine at a site about six miles northeast of Glorieta. Included would be a power generation plant to provide power for both the mine and the mill.

chapter 17

✲ ✲

A Town In Its Infancy

Next morning, the sun did not appear until after nine. Because the mountains were so precipitous on both sides of the river, the meadow was in deep shadow most of each day. Even during the longest days, it received not more than five hours of sunshine. The outside temperature that first morning was in the low thirties. When the sun did shine, the air began to warm, and only then did Mom let us go outside. She had not expected such cold weather; our heavy clothes had been packed in a trunk that had not yet arrived.

When we did venture out, we were surprised that we were not the only residents of the meadow. A doe and her fawn grazed contentedly on clover, chipmunks scampered about, and bushy-tailed gray squirrels chattered at us from lofty pines. Steller's jays scolded us for invading their territory, while robins pulled earthworms from damp soil near the river. Hummingbirds, butterflies, and bumblebees flitted from one early blooming wildflower to another.

Our source of water was a small pool on the river which Dad had enclosed with river boulders to still the turbulence of the rushing water. At the edge of the pool, he had used other boulders to construct a low wall to prevent us from falling in when sent to fetch water. A small pail was tethered to a tree with a small rope. It was long enough to allow the pail to be thrown into the pool to draw up a full pail. The water was, in turn, poured into a larger pail used to carry it to the tent. He showed us how it worked, for Mom would be sending us to fetch water often. Because we were quite small, LaVoyse and I had to fetch as a team.

While at the river, we saw water ouzels, more commonly called dippers, and a belted kingfisher—one bobbing dipper stood on a submerged rock with water rushing over its feet and the kingfisher perched on a tree limb just above the water, its eyes glued to something in the river. Suddenly, it dove in and came up with a small fingerling trout in its beak. Both birds fascinated me—they were far more interesting than the jays and the robins.

After lunch, the whole family went for a walk across the north bridge and up the road about three-quarters of a mile to the area where the mine was located. Eventually, the area would become a bustling small town, but that day there were only 18 existing buildings. In addition, there were several industrial buildings and structures under construction at the mine-site above the Pecos road. Four of the existing buildings (a mess hall and three eight-room bunkhouses) were relatively new, having been built under Dad's supervision. The remainder were the mine office, five smaller bunkhouses, and eight residences that had been built by Goodrich-Lockhart sometime between 1918 and 1921.[1] The eight residences would eventually house families of office personnel. Three were currently occupied by Chief Accountant, William Fisher; Chief Engineer, Eugene Anderson (the second man hired); and Mine Superintendent, James Coulter.[2] All were baching[3] while awaiting the arrival of their families. Chief Bookkeeper, Jimmie Russell (the first man hired), lived in a bunkhouse adjacent to the mine office

Words cannot adequately describe, nor graphically portray, the topography of the terrain upon which these buildings were situated. An aerial photo frontispiece and a map on the inside of the back-cover of this book, as well as other included photos and maps, will better portray the area.

Generally, however, the locale encompassed four distinct areas which I have delineated in cross section See Fig 14 and define as follows:

Area 1) A relatively flat area on the valley floor between the juncture of Willow Creek and the Pecos, and extending about one-half mile north.

Area 2) A much smaller, escarped flat area, about 20 feet above the valley floor, formed by alluvial deposits from Willow Creek and located between the valley floor and the mountains to the East.

Area 3) A small lower-portion of the relatively steep mountain north of Willow Creek where a prospector, J. J. Case, had discovered the ore outcrop, and had sunk what he called the Evangeline shaft. [4]

Area 4) A much larger lower-portion of the more gently sloped mountain south of Willow Creek.

---

1 The Pecos Corporation, a spin-off from Goodrich-Lockhart, had sold 51 percent of their interest in the mine to American Metal Company, Ltd. They in turn had purchased it in 1918, and had attempted to exploit it before they sold the 51 percent share. During that time period, they constructed these buildings.

2 Whenever known, given names or initials will be used at first mention of an individual, otherwise Mr. or Mrs. will be used.

3 Although *baching* is the correct and preferred spelling, *batching* is used by most people.

4 The top of this mountain was the first area upon which the sun shone in the morning, and unofficially became known as *Sunrise Peak.*

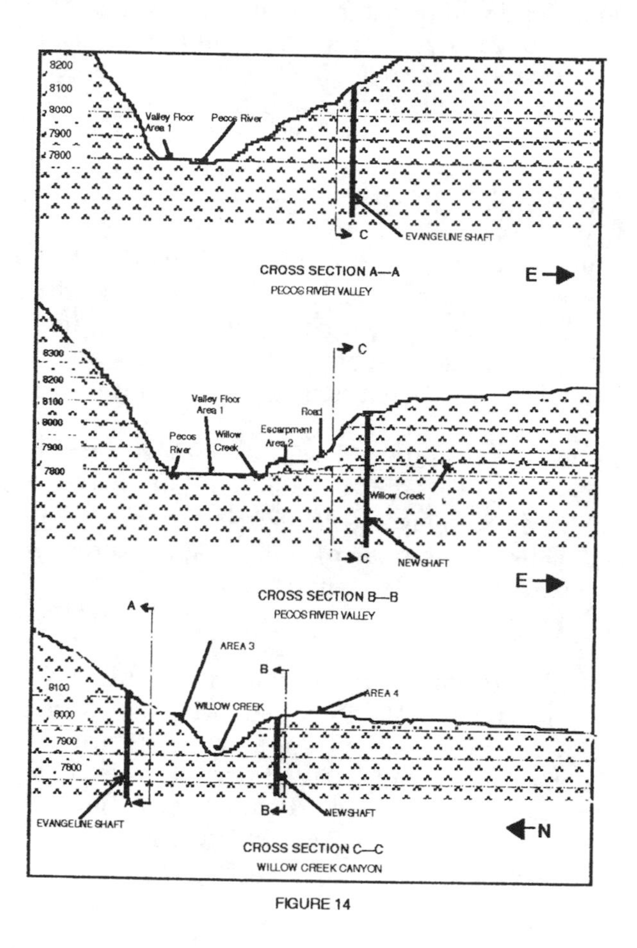

FIGURE 14

Above the escarped Area 2, in Area 4, the new mine complex was under construction. The valley floor was devoid of any permanent buildings except for three—the messhall located due west of the office that sat at the edge of the escarpment at the 90 degree bend in Willow Creek, and two bunkhouses. A half-dozen tents were scattered among the trees near the river bank; they were occupied by married men who anticipated bringing their families to the mine when a general store could provide the necessities for family living. The majority of the employees at that time were single and lived in the eight bunkhouses. A few married men also lived in them and would eventually move their families to the mine; all were fed in the messhall. Until that week, Dad had lived in the newest bunkhouse on the escarpment and had eaten in the messhall.

As the construction and mining activities became more intense, there would be a need to build more bunkhouses for the single men and houses for the married men. The three bunkhouses located closest to the office were used to house office and engineering personnel. The remaining five, clustered a little further away (three on the top of the escarped area, and two on the valley floor) were for both construction and mine blue-collar employees.

Of the eight existing residences, six were located north of Willow Creek on the west slope of Sunrise Peak overlooking the road leading north to the little town of Cowles. The other two were located about 300 yards south of Willow Creek, overlooking the road from Pecos.

In addition to these eight houses located in the vicinity of the mine, there were five others on the mine property somewhat more removed from the mine-site. Four were located near its southern end at Holy Ghost Canyon, and one was above the mine complex, about one quarter mile to the southeast. This had at one time been the Chapin ranch house with guest quarters, maid quarters, a large barn and two carriage houses.[5] The mine General Manager, J. B. Haffner, occupied this house with his wife. Besides Mom, she was, at that time, the only other woman at the mine.

Within the next couple of weeks, however, there would be several new families move to the mine and occupy the houses at the southern end. Near the mouth of Holy Ghost canyon, the Simmons Ranch would be occupied by the Company Doctor, Warren Smith, his wife Alma, and their five children, Maxine, George, Dan, Dick, and Lorraine. One of Dad's crews was just completing a small hospital addition to the main house.

H. Riley Littrell and his family, who would operate the dairy, moved into the house across the river from the cave, another Simmons Ranch house. He and his wife, Ella, had three children living at home, Ethel, Ruby, and Norman. A married son, Ralph, would move his wife and three children, Ralph Jr., Donald, and Jean, to a cabin across from the Nunn Place (now known as Tres Lagunas) about two miles south of the dairy. An older daughter, Gertrude, who was married to Herb Pritz would also move to the town.

Up Davis Creek from the Littrell's, Billy Montgomery would move his sister, Mary MacDonald,[6] and her two daughters, Doris and Jane, into the old Marlow Ranch house. The old Cornell house, located up the Pecos River about one quarter-mile from the Littrells, would be occupied a bit later by a mine foreman, Dan Mangum, his wife, Marie, their two sons, Owen and Ralph, and three daughters, Betty, Jayne, and Ruth. Another mine foreman, Elvin Fuller, would soon move his wife, Marie, and their family into the first of the six houses overlooking the Cowles road.

Between the mine-site and the Haffner house, adjacent to a trail connecting the two, was a gently sloping meadow at the upper end of a shallow vale. Four three-room houses were to be built on the high side of this meadow. The one nearest the Haffner house was reserved to be our future home. Dad took us by to see the location, and Mom was elated that we would be living in such a beautiful spot. On the other hand, she was somewhat apprehensive that we would be the closest neighbor of the top mine official. Completion of the four houses was not scheduled for another year. Meanwhile, when one of the two-room houses was completed, we would move into it.

When compared to our 10-foot by 14-foot tent-house, to Mom, the three-room house would be akin to a mansion, although it would be only 432 square feet in size. Eventually, over 100 company houses would be constructed, consisting of either two or three rooms, or in one instance, an adobe building with ten, one-room apartments.

---

5 When American Metal Company Limited decided to open the mine for production, the five former ranch houses were also purchased—the Chapin ranch house for the General manager's quarters; the Simmons ranch houses for a hospital, the doctor's quarters, the dairy and the dairy operator's quarters; and the Marlowe and Cornell ranch houses for general housing purposes.

6 For reasons never understood, it was later revealed that Billy and Mrs. MacDonald were not brother and sister, but husband and wife. Whether or not the two girls were only Mrs. MacDonald's or the offspring of both of them was never known. They had another daughter several years after moving to the mine, an event that triggered the revelation about their marital status.

There was yet no post office and no store. Mail was received addressed: John Doe, In Care Of Amco, Valley Ranch, New Mexico, and could be picked up at the mine office. The nearest groceries were available from Harrison's Country Store in Pecos. They would soon begin to run a delivery truck to the mine three days a week. Orders placed with the delivery man on one delivery day would be delivered the following delivery day.

When Dad had shown us all that there was to see, we stopped by the office to pick up mail. We met Mr. Fisher, whose wife would become one of Mom's good friends, Mr. Anderson, and Mr. Russell.[7] It could not have been known that day that Russell and I would be the last two of the original residents. Nor could it be known that he would drive me back to Glorieta a little over thirteen years later to catch Santa Fe's *El Capitan* to my new home in California. He would live in his Holy Ghost Canyon home for a few more months, then move to Santa Fe.

It had been a long day; we four children were all tired from the walking and climbing in the thin air of the 8,000-8,300 foot altitude. We would be ready for bed early that night. Mom carried LaVoyse, and Dad carried me back down the old road to our isolated tent home.

---

[7] Old and young alike seldom referred to him as Mr. Russell—he was just Jimmie Russell to everybody. It was assumed by all that he was about the same age as my father—28 to 30 in 1926. When I talked to him in 1992, I said to him, "You must be around 94 years old now." His reply, "No way, I'm only 87." When I reminded him that everybody seemed to think he was older, he said, "I know, but I lied about my age. They would have never hired me had they known I was only 21 at the time."

chapter 18

* *

The First Year

Life on our isolated meadow was a new and exciting experience for us children, but isolation and primitive living were not new to Mom. She married Dad in 1915 while she was still 14 and he was 21. A year later, they moved to New Mexico to stake out a homestead. About 20 miles east of Mountainair, they found 160 acres to their liking. On it they built a small, one-room house, they cleared and plowed the land, and around it they erected a barbed-wire fence. They bought a cow, a work-horse, and some chickens, and they waited to plant a crop of pinto beans the next spring.

Being short of funds, Dad found work as a carpenter in Mountainair. Riding back-and-forth 20 miles on horseback each day was out of the question. He had to stay in town during the week, coming home late Saturday night and leaving for work early Monday morning.

Homesteading laws required three years of uninterrupted occupancy. During that first year, at age 15 and pregnant with my oldest sister, Mom stayed on the homestead alone for six days each week for about nine months. She planted and raised a garden, and she tended and milked the cow. She hauled water from the nearest neighbor's well, about two miles away; she hoed the weeds from the growing bean crop.

She was alone with the responsibility of the homestead for around eight months out of each of the three years, and she bore two children while doing it. She was truly a pioneer woman. Now, in this new situation at the mine, Dad was home every night, groceries were delivered to the tent door, water was a few steps away, and there were four children to fetch it. This was a life of ease compared to her first few years of married life.

The only people we ever saw was the delivery man from Harrison's or those who came by fishing the river for the abundant trout. One day a family of three came by: Tom Stewart, his wife, Eloisa, and their teenage son, Marion. Mrs. Stewart was not fishing, so she struck up a conversation with Mom. This led to a long and lasting friendship between the two. Mr. Stewart, it turned out, had been one of the first rangers in the National Forest Service. He was now the Supervisor of the Pecos River Forest Reserve in which we lived. Mrs. Stewart was his second wife, somewhat younger than he. They lived on a small ranch near Cowles—Stewart Lake in the Pecos Wilderness is named for him.

As for us children, we needed no playmates—we had one another. When not playing, we were busy with Mom's assigned chores. She was obsessed with cleanliness.

We always had to be clean—both our bodies and our clothes. We carried hundreds of gallons of water and collected cords of firewood to meet her expectations, for there were clothes to be washed and baths to be taken. Both were accomplished in the same large washtub. Water was heated in another tub resting on three rocks, a wood fire blazing beneath.

Early one morning, LaVoyse and I were sent to fetch water. The tethered pail had been left in the river pool overnight, and had sunk to the bottom. I pulled it up and began pouring the water into the carrying pail. A small water snake had taken refuge in the pail during the night; I saw it only when I poured it from one pail to the other. Upon seeing it (to my young eyes it seemed to be a monster), I threw the carrying pail, snake, and the water into the river. I went screaming back to the tent; LaVoyse followed close behind, screaming as loud as I. The pail floated down the river never to be seen by us again.

The first magazine I ever remember seeing was a copy of the *National Geographic* someone had given to Dad. An article about an African tribe was one of the featured articles. Even at the tender age of five, I was shocked at seeing pictures of bare-breasted women.

Rainfall during the last two weeks of May and the first week of June totaled over four inches—slightly over one inch fell as snow. The run-off from the rain and melted snow at the higher elevations caused the river to rise about three feet, the turbulent flow creating an almost deafening roar. One night, the lower of the twin bridges partially washed away. From that day on, auto traffic on the old road was impossible.

The following morning, we awoke to the bellowing of a herd of milk cows huddled together at the approach to the bridge. Cows from Littrell's Dairy were permitted to roam and graze freely over the mine's property after the afternoon milking. Early most mornings, they would return to the dairy down the old road and across the twin bridges. This morning they were stranded, and they were loudly voicing their discontent. Dad had to go to work, but instructed Delta and me to try to get them to cross over what was left of the bridge; they simply would not go. Finally, one plunged into the rushing water below the bridge, and was quickly washed downstream. The others followed along, one-by-one, bellowing loudly as they too were washed away by the raging water. About two-hundred yards from the bridge they managed to regain their footing and crossed to the other side. Milking that day was very late. How vividly I remember that scene over 65 years later!

Mom became concerned that the upper bridge would also wash away and leave us stranded on the meadow. She managed to persuade Dad to seek another location, closer to the mine-site, and not so isolated. He chose a spot on the west side of the river almost directly west of the future post office. See ⊗ on Fig 13

As yet, on that side of the river, there were no other occupants on mine property, but adjacent to the Amco's holdings, on the north side, was real estate and a home owned by Louis Rivera. A one-lane road led to his property across that belonging to the mine, and bridged the river about 350 yards north of our new tent-site. This was the access road and bridge to our new home. The back of the tent abutted against a rock cliff, and 20 feet to the front, the river flowed. The valley floor was pinched off by the river flowing against the rock cliff just a few feet to the south of the tent. A heavy stand of willows and alders hid the rest of the town from our view, so we had almost as much privacy as we had between the twin bridges. It was only a short walk to what would one day be the post office, to the school that was being built on the valley floor near the confluence of the river and Willow Creek, and to the mine-site. Dad could easily walk home for the noon-time meal. Moving occurred one Sunday in late August, again in the company's Model "T" truck. Nobody would ever live on the meadow again.

Rivera owned a tame burro that would often graze on the lush vegetation growing near our tent. Dad fashioned a rope halter, and we children would often ride him, three at a time, around the meadow. Once Delta and I rode him to the bridge to meet Dad when he came home for lunch; LaVoyse ran along beside us. Dad picked her up to set her down behind me, but an open safety pin in her diaper pricked the burro. He bucked, throwing me over Delta's head, and I landed on the ground with a thump! Dad still had LaVoyse in his hands and was able to save her from my fate. Delta was far enough forward that he was spared my indignity. The burro made his point; I never rode him again.

Around the middle of July, a one-room school house had been started when it seemed probable that there would be no more than 35 pupils attending first through eighth grade. A month later, the number of pupils expected to attend had increased to 65, and a second room was rushed to completion. The county provided two teachers—Miss Parsons, the principal, who would also teach fifth through eighth grades, and Mrs. Clark, who would teach the first four grades. A small two-room house was built across Willow Creek from the school for Mrs. Clark and her son, Tom, age six. Another small one-room house had been constructed for Miss Parsons on the northwest side of the school. The attendance situation would change dramatically during the next two months. Meanwhile Balpha and Delta began school shortly after we moved, she in the fourth grade and he in the second. LaVoyse and I were too young to go and spent our time playing around home.

By mid-October, our two-room house was complete on the hillside across the river. We moved again on a Sunday, that being the only day, every fourth week, that Dad did not have to work. We boys helped him strike the tent and raze the pine-board floor. Our sisters helped Mom set up house-keeping in our new wood-frame house; tar-paper covered the outside walls, and beaver-board covered the inside walls. The tent had been our family's home for five months; for Delta and me, it would serve as our room again beginning about six months later.

Meanwhile, at the mine, Joseph Matson, Sr. had replaced Haffner as General Manager just two days after we had moved from Mountainair. His family did not come until around August 1.

Because of a continuing and serious problem of hiring employees who would stay on the job, a decision had been made by Matson to construct houses for about two-thirds of the required miners and all of the above-ground skilled employees, an additional total of 80-plus houses. In addition, other workers would be permitted to select a plot of ground on specified mine property upon which they could construct their own domiciles. Lumber, nails, and tarpaper would be provided by the company at cost. The overall plan proved to be the hoped-for solution, because the work force began to inch upward shortly after the plan was announced; employees knew that there would soon be housing for their families, or that they could build their own houses at a nominal cost.

Fifteen new houses were started and in various stages of construction by the end of summer when we moved to the tent-house across the river, and construction on 35 more was about to begin. Seniority would determine who got the houses that would rent for $4.00 per month for a two-room and $6.00 for a three-room house; electricity would be provided at no cost.[1] The average construction cost per room was about $250 for a total of around $50,000. This expenditure had never been anticipated by Amco Ltd. management; only $12,500 had been budgeted for construction of bunkhouses for single men, nothing for houses for married men. In essence, it was never anticipated that there would be many families living at the mine.

In addition to the company houses, employees with little seniority began constructing their own places, for the most part tent-houses later to be converted to frame houses. The burgeoning increase in population required a building to handle the increasing flow of mail being delivered out of the mine office. Mail still had to be addressed in care of the mine at Valley Ranch, since the mining town had no name. A building was completed in mid-December which housed not only the mail deliveries, but also a barbershop and a poolhall. In return for handling the mail, John Gould was given the barbershop and poolhall concessions at no cost. This would change in a very few months.

A 50,000-gallon water storage tank was completed in early June on the top of a hill above the change house. Water was taken out of Willow Creek about two miles upstream; its purity was over 99.8 percent, so no treatment was needed for domestic use. Initially, a distribution system was installed to serve the mine complex, Matson's quarters, and the area around the valley floor. The construction of the new houses required extensive lengthening of the distribution lines for both domestic water needs and fire protection. Each group of four houses was provided with one spigot from which water could be drawn. Two-inch fire hydrants were strategically located in a four-foot by four-foot shed with sufficient firehose to protect any house within 200 feet. As a practical matter, they would prove to be quite useless.

Throughout 1926, the availability of groceries was limited to four sources. Just north of the mine on the Cowles road, located in an old red building on Christino Rivera's property, an enterprising man had opened a small grocery store. Besides the few groceries and some dry goods, he sold white gasoline and kerosene for the lanterns everybody had to use for a light source. The white gasoline was the only item my parents ever purchased from him. For other groceries and supplies, they relied on Monday, Wednesday, and Friday deliveries from Harrison's Country store in Pecos

For a couple of months, they also subscribed to a produce service in Albuquerque, whereby each week a box of fresh produce was shipped at a fixed price per box. The customer had no choice in the selection provided, but there was quite a good variety of good quality fruits and vegetables. In late September, however, Mr. Gilcrease, who owned a small truck garden and grocery store on the Albuquerque highway about ten miles west of Santa Fe, began making deliveries of fresh produce to the mine each week.

---

[1] About three years later, it was discovered that many were heating with portable electric heaters. The result was that meters were installed and electricity was no longer free.. The cost was minimal, however, .6 ¢ per kWh, and an average electrical bill was less than $1.00 per month.

In exchange for three meals while he was in town selling his wares, Mom got much of her produce at no cost, and the Albuquerque produce subscription service was canceled.

Gilcrease sold his produce on a charge basis, and he kept those charges recorded in pencil on the wooden side-panels of his delivery truck. Anybody could examine the truck carefully and find out how much the neighbors owed him. As he was paid, he would simply cross out the charge and start anew, each customer having a designated spot on the panels for their charges. When he ran out of space, he would sandpaper away the figures and start afresh. He made the unfortunate mistake of trading in his old truck without copying the charges one week, and had to rely on the customers to tell him how much they owed.

Our adjacent neighbors when we moved to the two room house were the Zumachs and their teenage son Frank, a seventh grader. Mr. Zumach was a mine foreman and they occupied the first of the two-room houses to be completed. The parents spoke broken English, having migrated from one of the Balkan countries. Others came from the same general area of Europe.

Frank seemed to delight in playing with Delta and me. He taught us many games and various things to amuse ourselves. He taught me how to roll a small metal hoop, a pastime that I enjoyed for a couple of years. On another occasion, having found a large abandoned truck tire in which he could curl up inside, he would have Delta and me roll him, head-over-heels, along the road above our houses. The slope of the road in front of the houses was minimal, so we could easily control the direction and speed of the tire. Closer to town, however, the road steepened, and when we reached that stretch one evening, we could not stop him; away he went, totally out of control. He managed to extricate himself just before the tire went off the road. Down the steep hillside it bounced again and again, finally landing squarely on top of an Hispanic laborer's tent, who happened to be home sick in bed. It tore through the tent's roof and came to rest on the sleeping man's bed.

I thought it best to beat a hasty retreat; Delta and Frank stood watching in awe as it bounded into the tent. The man came rushing out in his pajamas, spotted the two boys on the road above, and took off in hot pursuit. As Mom told it, they rushed into their respective houses with faces white as sheets. He arrived at the Zumach house first and began pounding on the door. All the while, he shouted something in unintelligible dialect; probably that he would like to kill two Anglo boys.

Mrs. Zumach, not knowing what circumstance elicited such anger, came out of the house with a rolling pin held high, shouting back, "You no touch my Frankie." The poor man fled, with her in hot pursuit. Later, when the Zumachs learned the full details of the incident, they purchased the man a new tent. Needless to say, that form of entertainment was banned.

An "Erector" set was my Christmas gift that year. Frank came over many times to help me construct all manner of mechanical devices. He had a keen mechanical aptitude, and probably instilled in me some of his love for mechanical things. His father died unexpectedly a few months after Christmas, and he and his mother moved to Las Vegas.

chapter 19

* *

The Town Is Named—A Friendship Blossoms

Around the beginning of March, 1927, a decision was made to select a name for the mining camp and to petition the Post Office Department to officially recognize the name as a mailing address. The residents were all asked to suggest names from which a final selection could be made. Although a number of names were submitted, the list was narrowed down to two. The Chief Accountant's wife, Mrs. Fisher, suggested the name Amco, a name that a number of people were already unofficially using. Eloisa Stewart, wife of the Forest Supervisor, submitted the name Tererro. Although the Stewarts did not live in the mining camp, they would receive their mail there when a post office was selected, so her selection was considered along with the others. At the time, their mail was addressed to a rural route number and delivered b y "Shorty" Gallegos, who made all the rural deliveries in the Pecos Canyon on horseback.

In any event, Matson made the final selection and chose Tererro. The origin of the name and its spelling have long been a matter of conjecture by many who did not live there in those early days.[1] I can vouch for the truth of the following account because I heard my mother, who was a close personal friend of both Mrs. Stewart and Mrs. Fisher, relate the story to numerous people, and because Mr. Matson verified it for me many years after the mine closed.

About one mile south of Willow Creek, just to the west of the Pecos road, there is a low depression in the terrain which collected water from rains and snow-melt. As the water evaporated, a bog was created.See *, Fig. 13 In it grew skunk cabbage and other moisture-loving plants. For some reason people found it convenient to throw rubbish into these weeds and, eventually, somewhat of a rubbish dump was created. Because one of the meanings of the Spanish word *terrero* is dump, the local Hispanics began calling it the *terrero*. When the water evaporated from the bog, salt deposits were left around

---

[1] In the book, *Trail Guide to the Geology of the Upper Pecos*, the authors speculate that the name was chosen because of the large dump of gangue (waste rock) which now exists in the area. When the name was selected, the dump simply did not exist, although there was a smaller dump on the north side of Willow Creek where Goodrich-Lockhart had disposed of the gangue from their development work. The Postal Service still adheres to the original spelling that it approved in 1927, **Tererro**, but most maps and the State of New Mexico use the correct Spanish spelling, **Terrero**. Still another book, ***Spanish** Place Names of New Mexico*, spells it **Terrerro**. The USGS map of the area uses both spellings. On two pieces of more recent correspondence, I have seen it spelled **Torerro**

the edges and each evening about dusk, deer would come out of the forest to lick the salt. The local Anglo people began calling the area a *deerlick*.[2]

Mrs. Stewart, thinking that *terrero* and *deerlick* were synonymous, and that the Spanish name had a poetic, imaginative ring to it, chose Tererro as her selection. However she misspelled the name by placing the double "r" at the end rather than at the beginning of the name. As a consequence of this understandable error, many of the early Anglo residents believed that the two words were synonymous; we liked to think that we lived in a town meaning deerlick, a word that is not even defined in Webster's Unabridged Dictionary. Salt lick is the definitive description.

On April 5, 1927, the Post Office Department notified Matson that Tererro would be the official address of the new post office and requested that somebody be recommended for postmaster so that one could be appointed forthwith. The name of Jim Welch was submitted and he was appointed to the position on April 15, by the then Postmaster General, Harry New.

Welch was a friend of Amco's president, Mr. Steele, having worked at an Amco Ltd. mine in Cuba. He contracted TB there and doctors suggested he move to a drier climate in Arizona or New Mexico. Steele asked Matson to find him a position as this seemed to be the ideal spot for him. He also took over the barbershop and poolhall concessions from John Gould, the individual to whom they had been given when the building was completed. Welsh had to purchase the equipment in the concessions from Gould at a cost of $1,800, a sum which had to be advanced to him by the company, since he arrived at the mine short of funds.

Shortly after the town was named, my parents instructed Balpha, Delta, and me not to return home from school during the noon hour; instead we were to meet Dad at the foot of the stairway leading up to the mine from the escarpment. They told us only that it would be a surprise; we could not imagine what it would be.

We met him as instructed and he led us up the stairway (134 steps and 11 landings in all). We walked through the mine complex and beyond on a narrow trail that began at the mine warehouse. We walked through a dense woods interspersed with Douglas fir, ponderosa pine, Engleman spruce, aspen, and scrub oak. In approximately one-quarter mile, we exited the woods at the lower point of an upward-sloping, triangular meadow. On the high side of the meadow were four houses, one completed and three nearing completion. We had been here before on our second day in Tererro. The completed house was on the site where Dad had told us we would eventually live.

Our parents had moved during the morning hours and Mom had our noon meal waiting for us. This was not a light lunch, for we always had dinner at noon, and a lighter meal (supper) in the evening. We would now be living in a house that was half again as big as the two room house—an absolute castle in our young eyes. As an extra bonus, we had running water on the back porch and would never again have to fetch water from distant faucets or the river.

I had made friends at school with Joe Matson, Jr., and now that we would be living as closest neighbors, I walked home with him that afternoon That walk home began a

[2] Today, most people seem to call it the frog pond.

friendship that was to last a lifetime. From that day forward, little happened to me for the next twelve years in which Joe was not a partner. Most of my long-cherished memories of Tererro involve incidents in which he and I were participants, and more often than not, it was just the two of us.

Joe was my age, had a half-sister, Isabel, who was fourteen, and a younger sister, Alice, who was three. His father was uncommonly personable, liked by everybody, and affectionately called "The Skipper" by mine employees when they referred to him. He always wore a hat outdoors, and in later years, also indoors.[2] He generally wore breeches and high-top, lace-up boots. He smoked a pipe which was usually in his mouth, whether lit or not. When he conversed with you, he would end each segment of his part of the conversation by taking three or four quick puffs on his pipe, then quickly saying, "Hm, hm, hm?" before waiting for the other person's reply. I always held him in the highest esteem; he became my role model and almost a second father to me.

Mrs. Matson, in contrast, was aloof, self-effacing, always very dignified and often in poor health. She rarely entertained; her only close friend was Mrs. Fisher, although she occasionally associated with Mrs. Smith, the doctor's wife, and with Mrs. Bemis from the mill. She intensely disliked living in Tererro and, in later years, took Joe and Alice to live the school year in Denver's Olin Hotel. She did not get along well with her step-daughter, Isabel.

Because of my close association with Joe, I probably came to know her, in a general sense, better than anyone outside her own family. I stood a little in awe of her, as did everybody else. Although we were her closest neighbors, Mom was in her house only a couple of times and she never came to ours. Mr. Matson was there only once, as I recall, and that was when Joe and I were in deep trouble, an incident to be told in another chapter. In 1928, the company sent the Fishers to Africa, and her one close friend moved away. She never had another in Tererro, although there may have been friends in Denver. Joe never mentioned them.

The Matson house faced a flat meadow about two acres in size. Slightly to the front and to the left side was the guest house and to the right rear, the servants' quarters. The previous summer the guest house had been enlarged by the addition of a bedroom and a bath. This had been necessary to provide quarters for the numerous V.I.P.s from the New York office or other Amco Ltd. operations. The main house, this guest house, the doctor's house, and the hospital were the only buildings provided with indoor plumbing.

In addition to the three houses, there were vegetable and flower gardens, a large expanse of lawn, and a root cellar. All of these were enclosed within a hog wire fence topped with barbed wire. Outside the fence, slightly uphill from the house, were two garages, one for the Matson's car and the other for the servants'. One had previously been a carriage house, the other a tack house. A large barn with horse stables was situated even farther up the hill, and behind it was a fenced pasture of about four acres.

---

[2] The last time I visited with him he was 97 years old. I walked into his room where he was napping on his side, his hat covering his head. I awakened him and he sat upright with a start, immediately placing his hat straight on his head. He asked who I was, and when I told him, he began asking sharp and knowledgeable questions about my family. He died a few months later and was buried in the National Cemetery in Santa Fe.

The company provided the Matsons a live-in housekeeper/cook, and a handyman who maintained the grounds, kept a supply of coal in the boiler room, did the gardening, and cared for the two company-owned horses, Nebbie and Blackie, stabled in the barn and pasture. The first two people to fill these two positions were a Black couple, Bertha Slaughter and her husband (his first name has been forgotten by Isabel, the only person living who remembers they were there). They had a daughter, Ben Ethel, about Isabel's age, who became one of Isabel's good buddies. The Slaughters quit after about nine months and another black couple, Louise and Milton Henson, were hired for the positions. Milton was the nephew of Matthew Henson, the Black sailor who, as his personal servant, accompanied Admiral Peary to the North Pole.

Except for a teacher who was of Black and Anglo parentage, Miss Cavey, no other Blacks ever lived in Tererro. The Hensons and Miss Cavey were the only Black people I knew until I moved to California in 1939. It was mainly through them that both Joe and I learned tolerance of other races. My own parents were often quite intolerant in their references to Blacks, but Mom did seem to relate well with these three.

Early on, Milton suggested that he would be willing to take Joe and me out on an overnight camp-out when the summer school vacation began. From that beginning, it became a weekly custom for him to take us, and occasionally several other boys, on hikes, wiener roasts, sleigh rides in the winter, or simply to teach us how to make sling shots, and bows and arrows, how to sharpen knives, to use lard to waterproof our boots, and many other outdoor skills. He was under no obligation to do so, but did it simply because he wanted to. To Joe and me, he could do no wrong.

Shortly after moving to our new house, the three adjoining houses were completed and occupied. Next door to us were the Sloans, then the Fergusons, and then the Gilberts. "Red" Sloan was electrical foreman, Ferguson was the tramway night-shift foreman (later to become general foreman), and "Baldy" Gilbert was night-shift mine foreman. None of them had children our age. Gilbert's severely epileptic half-brother, Charlie, lived with them and often had seizures in our presence. We were told to let him lie and he would recover in due time. They also had a son, Bill, about age 20, who would soon marry and move to a house of his own. The Fergusons, Harold and Bea, had a daughter, Lela Bea, shortly after moving into their house, and several years later, a son, Harold.

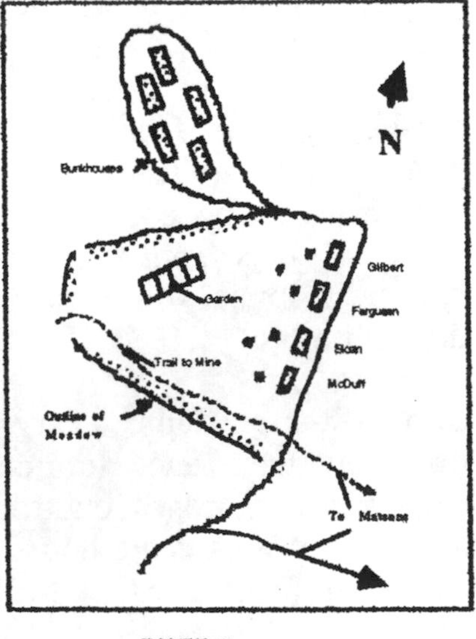

FIGURE 16

Below the houses to the north side of the meadow, there was a fenced garden for use by all occupants of the four houses. Each was allocated one-fourth of the space; ours was the lowest and Gilbert's the highest. Up the hillside from the garden, five new eight-man bunkhouses had been completed and occupied about the same time as the houses. See Fig. 15 These nine structures, along with the Matsons', were the most removed and isolated of any in the community, and we had a great deal of privacy not enjoyed by those living in the high density houses elsewhere.

Louise Henson was an excellent cook, and had volunteered to teach Mom how to bake bread and rolls. One Saturday in June, 1927, while she was helping Mom, the Sloan house next door caught fire. There being no telephones, Joe was sent running to get Milton and I was sent running to the mine to get help. Mom and Louise used our garden hose to spray water on the tarpaper on the end of our house; it was beginning to scorch from the heat. Mrs. Gilbert and Mrs. Ferguson sprayed the end of the Ferguson house.

Milton arrived and dragged the fire hose from its shed near our house, but he was too late; the fire had burned out of control. By the time other help arrived from the mine the burning house was totally destroyed. The women's efforts saved the adjoining houses from also being engulfed in flames, but the tarpaper had to be replaced. This was the first of several totally destructive house fires over the next twelve years. In the future, the fire hoses proved useful only in preventing nearby houses from burning.

After the Hensons quit, the housekeeper/cook position was filled by Jessie Gonzales; her husband, Fidel, or "Chief" as he was always called, worked at the mine. They were of Spanish extraction. The handyman position was filled by Shorty Gallegos, the former horseback-riding, rural postman. He also was assigned janitorial chores at the mine office.

Shorty's new work was far less strenuous than his previous mail-delivery job on horseback. He described to Joe and me how, on many cold winter mornings, he left Pecos at dawn to deliver the mail between Pecos and Cowles. His ride might be during or after a heavy snowfall which slowed his progress to a snail's pace. After having ridden over 40 miles, delivering the mail on his outbound ride, he would arrive back home in Pecos after midnight. Weary and cold, he would care for his horse, eat a hot meal, catch a few hours sleep, and then do the same thing over again the following day. He, and also his horse, had truly epitomized the true essence of the postal motto, *"Neither snow, nor rain, nor heat, nor dark of night, stays these couriers from the swift completion of their appointed rounds."*

chapter 20

* *

School Days

The ever increasing number of families moving to the mine during the last four months of 1926 increased dramatically the need for additional school space. By November 10, there were 95 students enrolled as compared to the 65 students at the beginning of the school year. To accommodate the increase, a part of the nearby messhall had to be partitioned off for use as a classroom until an addition could be built on to the school house. The County was asked to provide another teacher. There was a problem, however; the County would provide and pay a teacher only for a class size of 32 or more students. At 95 enrollees, one class would have only 31 students. The County agreed to provide a teacher, but for lack of a single student, Amco would have to pay the salary.

To avoid this, Matson, who was for all practical purposes the local school board, made a decision to permit any child who was five years of age to start school in the first grade. Four qualified, and I was one of the four. The others were Joe Matson, Jr., Jane MacDonald, and Dick Smith. Dick was the youngest, having turned five on October 20. The enrollment was now 99. Four of us started to school two months late, but a full school-year ahead of schedule.

The addition to the school made it into a "T" shape. The hallway/cloakroom at the intersection served all three rooms. Each room was 16-feet by 32-feet and the hall/cloakroom, where they all joined, was 16-feet square.See Fig. 16 The addition was completed just before Christmas, along with an addition to the teacher's quarters in which Miss Parsons lived. The new teacher, Miss Cavey, would live in this new room. Miss Parsons would teach the seventh and eighth grades, Mrs. Clark the fourth, fifth, and sixth grades, and Miss Cavey the first, second, and third grades. The converted room in the messhall was returned to its former use; this would not last long, for the school enrollment continued to increase as new families moved in.

Miss Cavey was a very personable young woman, perhaps about twenty years old, and I, as a five year old, was very taken with her. Miss Parsons, on the other hand, seemed old and a grouch. She was a rough, tough individual, and a strict disciplinarian. She came to school wearing cowboy boots and hat; she could ride a horse with the best of the men. She had two strict rules at school—snowballs could not be thrown, nor could you wade in the Pecos River on the west side of the school, or in Willow Creek on the east side. Despite the restriction against throwing snowballs, some of the older boys did throw some at the girls one day. Miss Parsons announced that every boy would get a licking after school that day even though most of us were innocent of the infraction.

At the appointed time, we all lined up, over 50 of us. The big boys insisted that the smaller ones go to the head of the line, and we foolishly obeyed—we were not wise to their strategy. Each of us received about a half-dozen good whacks with her belt. Near the end, when she came to the bigger boys, Miss Parson's arm began to tire and the whacks became less severe. Roy Lynch, the biggest boy of all—one of the culprits who threw the snowballs—was last in line and bragged that he barely felt the whacks. The older boys had started it all, but the younger ones received the brunt of the punishment. However, the punishment worked—as long as Miss Parsons was principal, snowballs were never thrown on the school grounds again.

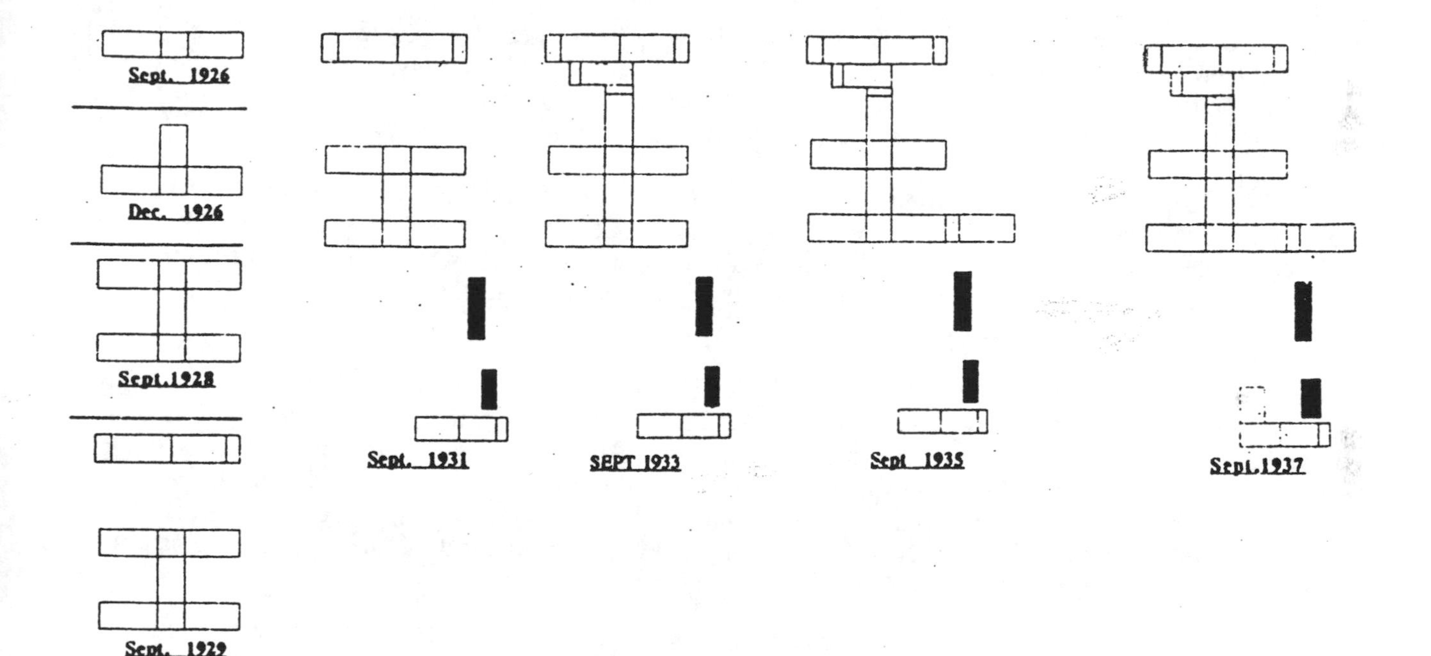

FIGURE 16

School House Configurations—1926 to 1937

Indicates Classrooms
Indicates Teacher Quarters

Note: Although they existed, teacher's quarters are not shown for years 1926, 1928, and 1929. Their size was proportionate to the size of the school

To enforce her rule against wading, she held unannounced foot inspections during recess. I could never recall why, but my feet were wet one day during one of her inspections. I might have passed muster had not Jane McDonald exclaimed for all to hear, "Leon, your feet are wet!" I was the only one ever to receive punishment for that infraction. I had to shed my pants in her classroom and stand behind the wood stove until my shoes and pants were dry. It was the ultimate indignity, for Isabel, Joe's sister, was in her seventh and eighth grade classroom. I was terribly embarrassed to take my pants off in her presence. To add insult to injury, I also received a spanking at home. The rule there was: get in trouble at school and you are in trouble at home. Dad always dealt out that kind of punishment with his razor strop, and it hurt. Much to everyone's delight, Miss Parsons moved to the mill for the school year starting September, 1929.

In the first year of school, Miss Cavey had the six of us in the first grade put on a short play entitled *Tom Thumb's Wedding.* Dick Smith was Tom Thumb, Jane MacDonald was the bride, I was the bride's father, and Joe Matson was the preacher. I was to kiss the bride after leading her down the aisle, and before turning her over to her betrothed. Someone had engaged a photographer to take pictures; the first and only picture he took was of that kiss. Lighting for the picture was provided by the ignition of a small amount of magnesium powder. When it ignited with its smoke and bright flash, all of us in the play became so frightened we began to cry, and refused to continue. No amount of urging by our parents could get any one of us to change our minds—Tom Thumb was not to be wed that night.

The same year, Mrs. Clark organized and presented a pageant in which parents and children alike were involved, Longfellow's *Hiawatha.* She was extremely interested in Indian culture, and obviously believed that this pageant would be a good way to bring some of that culture to the somewhat uncultured mining community.

The pageant was presented, complete with a teepee village, on the hillside and flat area between Mrs. Clark's house and the doctor's office. See photo page xiv right corner. The audience sat on blankets on the school grounds on the opposite side of Willow Creek. None of the participants had speaking parts; they simply acted out their parts as Mrs. Clark read the poem for all to hear.

In the pageant, Mom played the part of Nokomis, Joe Matson played the young child Hiawatha, Isabel played Minnehaha, and Marion Stewart was the young man, Hiawatha. Many others, whose names I can't recall, were also involved. I don't know how it affected others, but Hiawatha has always remained one of my favorite pieces of poetry because that pageant made it real in a little boy's mind. After teaching four years in the Tererro school, Mrs. Clark taught at the mill and then accepted a teaching position at the Indian school in Santa Fe.

One morning in the spring of 1927, fifth-grader Maxine, Dr. Smith's eldest daughter, took a two-ounce bottle of ether from her dad's office to school for some experiment she had in mind. Today she doesn't remember what that experiment was. She left the bottle of ether in her coat pocket in the cloakroom; the closure was not on tight. Ether fumes began seeping into the classrooms, particularly Mrs. Clark's, whose desk was close to the cloakroom doorway. She became faint and sent a student to fetch Miss Parsons. As Miss Parsons walked through the cloakroom, she realized that an odor was emanating from that room and her search revealed the bottle in Maxine's coat pocket.

She determined to whom the coat belonged and asked Maxine what was in the bottle. When told it was ether, she promptly led Maxine the short distance up the hill to her dad's office crowded with sick folks. He said he was too busy to discuss the problem with Miss Parsons at the time, but told her to take Maxine back to school and he would take care of her at home that night. Maxine wouldn't say what happened, but whatever it was, she surely learned never, ever to take that kind of experiment to school again.

The average school attendance in 1927 was 125, and in 1928, it had increased to 164. Two additional classrooms had to be constructed that year, making a total of five. A poll taken in early August, 1929, showed a potential enrollment increase to 227 at the beginning of the school year. Two more classrooms were begun on August 15 to meet the increased demand. There would now be seven rooms for eight grades. See Fig. 16 An additional room also had to be added to the teacher's quarters to provide room for two more single teachers.

The County Board of Education also agreed to provide teachers of Matson's choice. Previously he had no say in who was hired. Miss Parsons was moved to the mill school and Miss Phillips, a more genteel and compassionate principal, was hired.

When considering the overall student body, Anglo children were a minority in the school. The predominant ethnic group, as it was throughout the town, was Hispanic in the lower grades. As they advanced in school, however, they began to drop out; very few stayed for the full eight years or, when the ninth and tenth grades were added, for the full ten years. As a result, the upper grades were predominantly Anglo. In 1935, the last year Delta and I attended before we went away to high school, there was a total of 12 in the 9th and 10th grades. Of these, 6 were Anglo boys, 3 were Anglo girls, and 3 were Hispanic girls. Of those 12, only 5 completed high school—4 boys and 1 girl.

The majority of those who completed the eighth grade simply dropped out of school, but a few parents who could afford to do so and who took education seriously, sent their children to school in cities or towns large enough to have high schools. Isabel Matson was sent to the Prep School of the Colorado Women's College in Denver. Joe attended the New Mexico Military Institute in Roswell. Others, such as Dr. Smith's children, attended Loretto Academy (a girls' school) and St. Michaels' (a boys' school) in Santa Fe. Some went to Santa Fe or Las Vegas High Schools, and a very few of the Hispanics went to Pecos. The number who attended high school, however, was relatively small—never more than one-third of those who finished the eighth grade—perhaps one or two from each class.

Although our mother had only completed the fifth grade and our father only the third, Mom was determined that we would have the best education possible. She became friends with all of our teachers; some became her best friends. We were taught that, other than our parents, our teachers were the most important adults in our lives, and that we should honor, respect, and obey them as if they were our parents; woe be unto us if we did not.

When my sister, Balpha, completed the eighth grade, she was sent to Mountainair to live with friends of my parents while attending high school as a freshman. For her sophomore year, she lived with my dad's mother in Oklahoma. As a junior, she attended a boarding school in Montezuma, a few miles north of Las Vegas, but fell in love with a young man from the mine, Edwin Adlon, dropped out of school, and got married. She was still 16, but since Mom had been only 14 when she married, my parents couldn't object too strenuously.

When Delta finished the tenth grade in 1935, Mom rented a house owned by the Littrells in Santa Fe, and while Dad bached at home, we went to school there, Delta and I to high school and La Voyse to junior high. Norman Littrell roomed and boarded with us so that he too could attend high school; this helped defray most of the cost of renting the house from his parents. The following year, only Delta and I stayed in Santa Fe, baching with an uncle who had just moved there. I attended my senior year while living with the uncle and his family that he had moved to Santa Fe during the summer. Mom returned to Santa Fe with LaVoyse for her junior year in 1938-39 while I stayed home and cooked for Dad.

When in Santa Fe, I tried to go home to Tererro every weekend and was always home over any holiday break. It was not easy. On Friday afternoons, I rode with Reuben Hesch who delivered the *Santa Fe New Mexican* to paper boys at the mill, Pecos, and Tererro. It was far more difficult to find a ride back to Santa Fe on Sunday. I made few friends in Santa Fe, since I was seldom there during times when I could have been socializing.

The Anglo boys, as a general rule, did not get along well with the Hispanic boys, and there was a lot of fighting between the two; girls got along better. Joe and I were such good friends that we basically ignored both the Anglos and the Hispanics, and never became embroiled in any of the fights. For one thing, we both knew better—our dads would have given us a licking if we had, and we had enough such punishment as it was.

Laurena Tucker, who was not very tall, got into a fight one day with Rosa, a rather tall Hispanic girl. Rosa had a long pony tail which Laurena grabbed and swung around Rosa in Maypole fashion. This somewhat evened up the difference in their sizes. As with most of the fights, this one turned out to be somewhat of a draw, since it was stopped by a teacher. Omer, Laurena's brother, probably engaged in more fights than any other boy, but the Hurley boys, the Jenkins boys, Roy Lynch, and Pat O'Hara had their fair share also.

Mrs. Sarah K. Ellis, oldest of the teachers, taught the seventh and eighth grades for a few years and then only the eighth grade after 1934. Because she was obviously several years older than the other teachers, we all believed she must be at least 60. In reality, she was probably in her early 40s.

She usually wore a gray wool, cardigan sweater while teaching, and in one of its pockets she carried a short length of small rubber tubing. If she walked up behind a boy who was misbehaving in some manner, she would whip out the rubber hose and lay on the shoulders of the miscreant a few good whacks. As a result, she was dubbed and referred to as "the old rip." She was the strictest disciplinarian in the school.

On one occasion, Omer, who was in the seventh grade, was caught in some act of misdeed, but ducked away from the intended blows with the dreaded tubing, slid out the other side of his desk, and began running through the aisles with Mrs. Ellis in hot pursuit. He finally ran out the door and headed for home. Knowing that Mrs. Ellis would contact his dad, he thought of a scheme to rationalize his having left school early, and to put the blame on Mrs. Ellis. He pulled out a handful of his own hair, then told his dad Mrs. Ellis had done it in school for whatever minor thing he was doing. It didn't work. She easily convinced Mr. Tucker of the true circumstances. From that point forward, Omer almost became a model student.

By 1931, the enrollment had increased to 278 students. To provide additional schoolrooms, the largest bunkhouse on the escarpment was converted to a new messhall, and the old one was converted into two classrooms, bringing the total to nine. In 1933, an increase of another 33 students required the addition of still two more rooms—one a classroom, and the other an office for Miss Phillips which included a small library. The 9th and 10th grades were added to the curriculum; Miss Phillips taught the 10th grade in her office. An additional 37 students in 1935, required yet another classroom, and in 1937, another room was added to the old messhall. This new room, and the two in the messhall, became the classrooms for the 8th, the 9th, and 10th grades, thereby separating them from the lower grades.See Figs. 16 The enrollment on the last day of school, May 26, 1939, was 358 students with 14 teachers, including Miss Phillips. In the high school class that year, there were 3, 10th-graders, 10, 9th-graders, and 14, 8th-graders. This was close to the ratio that had prevailed for several years. When I was in the 9th grade in 1934/35, there were 6 of us plus 5, 10th-graders and 16, 8th-graders. After 1928, each grade beyond the 2nd grade was smaller than the preceding one. By 1931, there were around 50 students in each of the 1st and 2nd grades with 2 classrooms required for each grade.

When the sawmill was moved to Indian Creek, a community of about 15 families sprang up; there were more or less 30 students that had to be bussed to school. For a bus, Ralph Littrell, at his own expense, built a wooden enclosure on a lumber truck that was not used during the winter of 1934. Children sat on three wooden benches built down the length of the truck bed. The smaller kids straddled the center bench. In 1935, the County awarded Ralph a contract to provide a regular school bus, and he purchased a used, 24-passenger, 1931 Model "A" Ford bus; he painted it a hideous black and orange.

The school was by no means equivalent to modern day schools, but I believe we had some of the finest teachers possible. We learned things that aren't emphasized today—geography, natural geography, ancient history, and penmanship to name a few. Marie Phillips, Sarah Ellis, Hilda Voetberg, and May Walter were four of the very best.

There were no organized sports activities in the school, but the boys had three favorite activities in which they indulged. In the fall, we usually played "shinny", a game similar to hockey. Instead of a puck, we used a small *Carnation* or *Pet* milk can, and the shinny stick (similar to a hockey club) was a length of scrub oak.

During the summer, we would all search for just the right scrub oak bush from which a long, straight branch, angled at about 30 degrees from the trunk, could be cut to include about four inches of the trunk. The twigs and bark were removed, the shank was smoothed, and the ends were rounded. After seasoning for a couple of months, we had a new, and hoped for, winning shinny stick. The name shinny derived from the fact that we all went around with skinned shins from being whacked on by an opponent.

Baseball was popular in the springtime. Since there were never enough boys to make two teams, we always played "work-up", sometimes called "rotation baseball." Play consisted of having two or three batters, and as many of the fielding positions filled as possible. As each batter struck, or fielded out, those in the field would move up one position, i.e., catcher to batter, pitcher to catcher, first base to pitcher, etc., and the losing batter would move out to right field. If someone caught a fly ball, that person and the batter exchanged positions immediately.

The other favorite game was "horse and rider" wars, a game in which the bigger boys would be the horses, and the smaller ones the riders. Riders would be held on the backs of the horses with their legs held firmly in the arms of those who were the horses. The object was for each rider to grab another rider and attempt to topple the opponents, both horse and rider, or dislodge the rider from the horse. Harvey Hurley was generally my horse and we were the victors quite often. Sometimes we played the game with the horse standing on stilts. Not only did he have to hang on to the rider, but also the stilt handles. To overcome this disadvantage, we started using duck stilts—stilts that had shortened handles strapped to one's legs.

In addition to the above, two other popular games were marbles and spinning tops. Harvey Hurley was the expert in both—he won the most marbles in most every game, and could accurately knock others' tops out of a circle drawn on the ground. He would sharpen the metal point on his, and could often split another's top into two pieces.

In nice weather, the girls played hopscotch, jacks, jump rope, and red rover. Red rover was a game in which a soft ball was thrown over the roof of the school building by a member of one team and retrieved by someone on the opposing team. That person would shout, "Red rover, red rover, let so-and-so come over." That person on the other team had to run around the building to the opposite side and attempt to get there without being hit by the ball thrown by the caller. If hit, the person became a member of the throwing team, the object being to eliminate everybody on the opposing team. The girls sometimes talked the boys into joining in the game.

In the winter, when there was fresh snow on the ground, boys and girls always played "fox and the geese" together, a game in which a large wheel with spokes, perhaps 15 to 20 feet in diameter, was trampled into the snow. The fox attempted to catch anybody trying to cut across from the outside rim to another part of it by running on one of the spokes. If caught, that person became the fox. For the younger children, there were swings, seesaws, and merry-go-rounds.

All of these activities were rather simple compared to what schools provide for students today, but we made the best of what we had. All in all, we had a lot of fun and we got a good, basic education, one which fitted us for our lives ahead—as parents, and for the majority, blue collar jobs; for a few, however, positions of responsibility in engineering, medicine, plant management, education, etc. The majority of the boys who became teenagers before 1935, found themselves in military service during WWII.

While the school attendance at the mine increased every year, that at the mill remained relatively constant. There were only two classrooms required throughout its life. Four grades were taught in each room; first through fourth and fifth through eighth. After eighth grade, students were bussed to Pecos to attend high school. The bus was a converted flatbed truck outfitted with two wooden benches located near the cab. This was covered in inclement weather with canvas much in the fashion of a covered wagon. Of the total of ten teachers who taught at the mill over the years, three were transferred from the mine school—Miss Parsons, Mrs. Clark, and Mrs. Walter. Mrs. Walter transferred in 1934, and remained until the school closed in 1939.

chapter 21

* *

A Melange of Unrelated Events

When we first moved into the new house, my sisters slept in the bedroom with Mom and Dad; my brother and I slept on a fold-out bed in the living room. Within a few weeks, however, Dad used the tent we had lived in at the two river home-sites to construct a tent-house about 25 feet down the hill from the main house. It had a pine board floor and three-foot high wooden walls. The tent flaps were replaced with a board-and-batten door. A sheet metal wood-burning stove provided heat during the cold winter months. This would be mine and Delta's "room" for the next 12 years. Our sisters took over our fold-out bed in the living room, and our parents now had the privacy of their own bedroom.

During the four or five months from late spring to early fall, the temperature ranged from the low 50's to the high 80's; living and sleeping in the tent-house was very pleasant during those months. As the weather got colder and colder, however, it became more unpleasant. In mid-winter, we sometimes had night-time temperatures as low as -32 degrees Fahrenheit. It was impossible to keep a fire burning all night. To keep warm, we had to sleep under four or five heavy quilts, our heads completely covered to avoid freezing our cheeks, noses, and ears. Heated bricks covered with old towels would be at our feet.

In the mornings, shivering uncontrollably and teeth chattering, we would dress as fast as possible, put on our fur-lined slippers, and run to the house. Dad would have a good fire ablaze in the living room stove. Our boots and gloves had been placed behind it the night before, so as to be warmed and ready for us to go outside for our morning chores. If we had been out in the snow the day before, they would have dried out overnight.

Upon returning from school each afternoon, Delta and I, as would almost every other boy in town, have to split the wood and kindling for the fires in both stoves—a daily and never-ending chore from early September to late May. If it had snowed the night before—it did often from early November to late March—our first chore would be to clear the steps and boardwalk between the house and our tent, as well as the dirt pathway to the outhouse.

Each of the four houses in our area had been provided with a six-foot by eight-foot storage shed; all of the ladies used these for their laundry rooms. Makeshift boilers had been fashioned from 55-gallon drums from which the top had been removed and the opening fitted with a small washtub. Mondays were always washday. Our first

Monday chore, therefore, was to fill the tub and build a fire to heat the water. Our water source was a hydrant on the small kitchen entry porch of each of the four houses; this was as close as we came to indoor plumbing, but it was far more convenient than the communal hydrants provided for all other houses. Since the hydrants and the riser pipes were exposed to the outdoor cold, quite often we would awaken to find the pipes frozen solid; on those days we would fire up a blowtorch with which to thaw them.

Lighting a blowtorch was not the safest thing for young boys to do, but Dad taught us early-on how to do so safely. The danger was in the need to heat the blowtorch burner with raw gasoline, which could easily be spilled on our gloves if not handled properly. Subsequently, when the torch was lit, the gasoline on our gloves could have caught fire and burned our hands severely. Fortunately, it never happened.

Having lit the torch, thawed the frozen pipe, filled the tub with water, laid a roaring wood fire under the tub, and perhaps cleared away the snow under clothes lines and other walkways, we could then have breakfast and get prepared to go to school. I envied Joe, for their house was heated by a coal-fired boiler that provided hot water and steam for radiant heat. He had no chores to perform, but did help me with mine after school. The mine offices, the doctor's office, the hospital, and Dr. Smith's residence were also steam heated—everybody else heated with wood, or in a few cases, with coal used in a wood-burning stove.

Kitchen stoves, for most families, were also wood-burning. Wood was the cheapest fuel available at $2.00 per cord. There were a few, however, who burned coal, which was several times more expensive than wood. There were two exceptions—the Matsons had an electric cook stove, and my parents had a stove fueled with gasoline which was purchased in 55-gallon drums. In these quantities, the cost was around 15¢ per gallon. The drum sat on the bottom landing of the steps leading up to the kitchen porch. Another of our daily chores was to transfer gasoline from the drum to the storage tank on the stove. This, like use of the blowtorches, was extremely dangerous, but we were taught how to handle it safely. Although this was a much more efficient cooking method, other families simply would not take the risk and stayed with their wood- or coal-burning stoves.

The stove's burners were supplied with gasoline from the storage tank, pressurized with a built-in hand pump which all of us, through the years, had to operate. However, to get a smokeless blue flame, it was first necessary to heat the burner supply lines so that gasoline flowing through them would be vaporized. This, as with a blowtorch, was accomplished by burning a small quantity of raw gasoline below the supply lines. This produced a yellow flame and lots of smoke which blackened the walls and ceiling. About every other year, when we were older, another of our hated chores was to wash and scrub both the walls and ceiling and repaint with *kalsomine*. We scrubbed the ceiling directly above the stove more often.

When the house next door was rebuilt, at a cost of $526, the Tuckers moved in as our new neighbors. Coulter, the mine superintendent had been transferred to Climax as general manager, Cliff Hoag had been promoted to Mine Superintendent and Mr. Tucker had been hired as Head Foreman. He also had four children, approximately our ages: Edwin, Twila, Omer, and Laurena. He was a widower, so his mother came to help raise the children.

He immediately built a tent house for his two boys when he saw what Dad had done. Joe was taken with the idea of also having a tent house, so the Matsons had Dad construct one for him, although Joe never slept in it as we did. Mr. Gilbert also built one for his half-brother. For about a year, our relationship with the Tuckers was quite cordial, but that would change dramatically. An incident between Joe and Omer would be the focus of that new relationship.

As we all came home from school one noontime, Omer and Joe had some disagreement and Omer shoved Joe. Joe in retaliation picked up a rock to throw at him, but before he could throw it, Grandma Tucker saw what he was about to do and called out, "Joe Matson, put down that rock." Joe's reply was, "I won't do it, you old b-----d." Grandma, in a rage, took off to tell Mr. Matson what had occurred and arrived at their house just as he drove up. In tears she had just finished telling him the situation when Joe arrived home. Mr. Matson asked Joe if he had called her a bad name to which he replied, "Yes and that's what she is."

Grandma then complained that Joe had been influenced by the McDuff boys and suggested that he not be allowed to associate with us anymore. Joe was made to apologize and Grandma was assured that Joe would no longer play at our house. She was not told that we would be asked to come to the Matsons' home to play until the whole thing blew over. Unfortunately, it never did. The Tucker children were forbidden to associate with us, and despite Mom's efforts to placate Grandma, she refused to give in.

Twila contracted spinal meningitis and was near death. Mom volunteered to help Grandma by cooking some meals for the family. She would have none of it. Because we used a dual outhouse, and the disease was contagious, we had to resort to using a chamber pot. Dad would carry it to a mine outhouse for disposal each morning.

Despite Grandma's efforts, we were all good friends when away from home. Mr. Tucker was left in a dilemma—if he overrode his mother, she might leave and he would have no one to care for the children, so we lived next door with an invisible wall between our houses. It was always necessary to make sure a Tucker was not in their side of the dual outhouse before we went to use ours, and vice versa. Mr. Tucker remarried a few years later, Grandma moved back to her home in Farmington, and peace returned to the neighborhood.

Shortly after the Tuckers moved in, the Hensons quit their jobs and moved away. There was a great deal of prejudice in the community against the Hispanics, but whereas that prejudice was subdued for the most part, the prejudice against the Hensons and Miss Cavey, who was part Black, was open and vociferous. Milton, who was quite black, could not go to town without taking some kind of verbal abuse. He stopped going.

Miss Cavey, who was of a lighter skin, felt some of the same abuse, so she found another job after only one year of teaching. The Hensons decided to leave after a white cross was painted on their front door and a note threatening them was left under the door. They reasoned, with good cause, that if they could not be left alone living next to the top mine official's home, they were not safe to continue living there at all. Although I was only seven years old at the time, the memory of their unfair treatment has been a major factor in my lifelong belief in the equality of all people.

The first Anglo child, a girl, to be born in Tererro arrived at about three one morning between May 9 and May 14, 1927. Her parents were Jeff Gibson, a miner, and his wife, Mildred, better known as Midge. Midge's sister, Jepp Reid, had come to be with them when Dorothy was born.

Midge had been rushed to the hospital by Jeff when she began having labor pains the previous afternoon. However, due to the accident which had occurred on the tramline on May 8,[1] all of the hospital beds were filled, so she was sent home and told to stay in bed. Dr. Smith told her that since this was her first child, she probably would still be in labor well into the next day, and that he would come by to see how she was doing the next morning.

However, Dorothy wouldn't wait until then. She decided to start to come into the world about two a.m. Jepp went to a nearby neighbor, an office worker, and asked that he go to the office to call Dr. Smith. With her considerable experience at having had four babies at home, Mom, a good friend of Midge's, had arranged to come to help when needed. Dr. Smith was supposed to pick her up on the way. However, he did not know that we had just moved into the three-room house and, finding our old house empty, continued on without her.

Meanwhile, the father panicked, decided he could not handle the situation, and ran off, leaving Jepp alone to cope. The baby popped out, but Jepp, having no notion of what to do, fainted dead away. Dr. Smith arrived to find the baby born, but not tended to, and Jepp stretched out on the floor. He quickly revived her with smelling salts, then took care of the new baby. (Mom went early the next morning to find out how things were going and found Jepp coping with a crying new baby).

Besides the numerous single babies born over the years, there were also one set of twins and one set of triplets. The twins were my first cousins, who were born at home with the assistance of a midwife. Since they were somewhat of a curiosity, Joe and I went to see them the second day—the first newborns he and I had ever seen. The triplets were born in the hospital, but were premature, and all died shortly after birth. Dr. Smith was quite disappointed that he could not save them, but he was faced with not having the right equipment for dealing with premature births. They were buried in the local cemetery, along with the only other three people ever buried there.

A very cold winter caused an epidemic of pneumonia in 1928/1929. There were over 50 cases; three adults and seven children succumbed to the illness. Several children died from diptheria the following year. This was a large ratio of the approximately 700 residents at that time. Dr. Smith was almost overwhelmed with care of the patients. Only the most severe cases could be handled in the hospital, due to a limited number of beds.

One of those who caught pneumonia was Jim Welch, the postmaster, whose lungs were already weakened by tuberculosis which he had picked up in Cuba. He partially recovered and made arrangements with Charles Earickson, a long-time resident of Pecos, to take over the poolhall and barbershop, and Matson requested the postmaster general to appoint him as the new postmaster. Welch moved to Tucson in the hope of recovering

---

[1] Jepp, who is now 90, says her niece was born on May 4, and that her sister could not be hospitalized because of the tramline accident. However, the mine records clearly show that the tramline accident occurred on May 8, so it must be assumed that the birth occurred several days later.

completely, but died a few weeks later. One of the children who died from diptheria was the youngest Anderson daughter, and she was one of the six buried in the local cemetary.

The pneumonia epidemic, coupled with the numerous accidents which occurred in 1928, resulting in the death of six men, led to a decision to increase the number of beds at the hospital. Three additional rooms were built during the summer of 1929, increasing the capacity to 12 beds. I would occupy one of those beds for nine days in 1930.

Many of us boys earned spending money caddying on the golf course, but my caddying career almost came to an end one August evening in 1930. I was caddying for Bud McInteer, the tramline inspector previously mentioned in Chapter 14. Hole #6 was a par three, and his first drive put him just a few inches off the edge of the green. He could see a birdie in the making if he could use his putter and sink his second shot. I smoothed the pathway, removed the sand from the cup, and stood holding the flag at the cup. In his fervor to sink the putt, he dubbed the ball and it rolled only a few inches, whereupon he threw his club in my direction. The butt end of the wooden shaft struck me on the left forehead, shattering a piece of skull about the size of a quarter into my brain, although I was not knocked unconscious.

I was taken to the hospital and lay on a gurney watching Dr. Smith and his nurse wife, Alma, prepare (both themselves and surgical instruments) for removal of the piece of skull. My father had been summoned and arrived before surgery began. Although he talked to me and tried to allay my fears of what was about to happen, I could not take my eyes off the doctor and his wife. I watched as he selected the surgical instruments and I watched Mrs. Smith put them in the autoclave. I watched while he took a silver quarter (all silver in those days) from his pocket and shaped it with a ball-peen hammer to fit the hole in my head—it was the only thing he could use for a silver plate. When shaped to his satisfaction, it too was placed in the autoclave, and during surgery, used to replace the portion of shattered skull he had to remove.

I was put to sleep with ether by Mrs. Smith as she had me count backward from 100—I think I made it to 93 before going to sleep. I awoke the next morning with a headache and nausea from the ether. It didn't interest me at the time that in all probability, I was the only person ever with a silver quarter permanently implanted in the skull. I have learned that Dr. Smith had used a silver dollar on a man kicked in the head by a horse.

During my stay in the hospital, my grandparents came to visit from Oklahoma, and my folks decided to take them to view the sights in Santa Fe. I was unhappy and cried because I could not go with the family on this very rare trip to the city. Miss Phillips, the new principal of the school, happened to stop by the hospital and learned of my unhappiness. She came to my room bringing with her a copy of *The Wizard of Oz* and read to me for about an hour. She returned every day to read more until I was released from the hospital. This endeared her to me and it was the beginning of a friendship which was to last for the remainder of her life.

The only problem the silver quarter has ever given me was that it was the cause of my rejection as a cadet in the Army Air Corps during WW II. Bud McInteer, on the other hand, was seriously affected by the event. Although he had been fearless as a tram rider, the strain and worry over his moment of anger—it could easily have been the cause of my death—drove him to become an alcoholic. In his sober moments, he married

Dolly Earickson, daughter of the new Tererro postmaster, but was soon divorced, and eventually ended up in the mental hospital in Las Vegas.

When Charlie Earickson became the poastmaster, Dolly worked with her father as postal clerk (she had been born with a severe hip problem and walked with a extreme limp.[2]), and her brothers, Johnny and Bill, operated the poolhall and the general store which their dad established. The store carried groceries, meat, clothing and hardware. There was also a small soda fountain of which Johnny was the principal soda-jerk. Bill Gilbert, son of the night mine foreman, was the principal butcher. Dolly, Johnny, Bill, and their father occupied a small house adjacent to the post office, while Mrs. Earickson remained in their larger Pecos home with their three smaller children, Larry, Frances, and Jerry. Another daughter, Mary Earickson Lunsford, lived at the mill.

The only public telephone in town, the old crank-type on a party line, was also in the store. All incoming or outgoing telephone calls for local residents were handled on this phone. The only other phones were in the mine office, the doctor's office, the hospital, and the Matsons' home. The single barber chair was manned by numerous barbers who came and went almost as frequently as some of the mine employees.

On cold winter days, the store, known as The Tererro Resort, was a favorite gathering place for those who came to pick up mail and buy a few groceries, particularly around 10:30 a.m. when the window opened for mail deliveries. In winter, the wood-burning, pot bellied stove was the focus of their attention while they waited. Customers often would go out to the woodpile to bring in a new load of fuel. For entertainment, other than gossip while they waited, there was a player-piano that cost a nickel per tune. For a while, there was also a couple of nickel slot machines.

Across the river on Louis Rivera's property, Emil and Henry Pick also opened a general merchandise store in 1929 at the opposite end of the meadow where we had lived in our second tent-home. Emil was the manager and lived in Tererro, as did their younger brother, Marcel, who also worked in the store. Henry lived in Santa Fe and, basically, oversaw management of two smaller stores they owned—one in Lamy and the other at Glorieta. However, on the fifth and twentieth of each month—payday—Henry would drive to Tererro with a bag of cash from their Santa Fe bank in order to have funds to cash mine employees' pay checks.

The store's location was not too convenient since all of the town's population lived to the south of the store, and for the most part, on the opposite side of the river. The Anglo population did not particularly care to patronize it, since that side of the river was inhabited totally by Hispanics and had come to be known by some as *Rivera Town,* and by others as *Chihuahua.* They built a much larger store in the downtown area in the summer of 1931, about 300 yards northwest of the Tererro Resort.

---

[2] At age 64, Dolly had surgery and the problem was corrected. However, after the many years of compensating for the defective leg, her spine had permanently deformed into an "S" shape which in turn caused her to have severe back problems after her leg was repaired.

H. S. Farley operated a much smaller general merchandise store on Rivera's property, just over the property line on the west side of the river. After prohibition was repealed, he also opened a bar in an attached building.

In 1932, when Franklin Roosevelt became President, and James Farley was appointed Postmaster General, the majority of postmasters throughout the United States were replaced with Democrats, that being the custom of that era since laws protecting civil service workers had not yet been established. Frank Papen, a friend of the Pick brothers, who also worked in the store, was appointed as the new postmaster in Earicksons stead. Earickson continued to operate his store for a few more years, but eventually sold his interest to the Picks.

The grocery and hardware operation was transferred to the big Picks' General Store, and a variety store was installed in its place. Albert Kahn, about 25 years old, the son of some of the Pick brothers' friends who owned a shoe store in Santa Fe, was hired as the variety store manager. Victor Martinez, son of one of the Hispanic miners, was hired as the chief soda-jerk. To replace Papen, another friend was hired, Fritz Warshauer

As I write this, it has been announced that the big Sears catalogue will be discontinued this year. It would have been difficult for people to get along without those Sears, Montgomery Ward, and Spiegel catalogues in Tererro—they served three purposes: First, merchandise could be ordered from them as long as they were current. Second, they were placed in outhouses to be used as toilet paper, even at the schools. Third, while concentrating in the outhouse, they provided a library with a plethora of ideas and dreams for things that one might someday be able to own. The only real toilet paper I ever saw, as a boy, was at the Matsons in their indoor bathroom.

When the Tucker boys, Delta, and I were around nine or ten years old, one night each week our fathers began taking us for showers to the mine's large change-house, and later to the supervisor's change-house on the north end of the warehouse. It was a welcome change to the tub-baths we had been accustomed to for the past four or five years. Most of the supervisory employees were permitted to, and did bring their sons to one of the change houses for showers. Unfortunately, the girls in town had to be content with washtub-baths.

Some of the early residents had heard a legend about the cave in the mountainside near Littrell's Dairy. As the legend went, every seventh year on the night of the seventh moon, seven Indians would enter the cave at midnight. The tribe's chief would be one of those who entered, but would not come back when six exited sometime later in the early morning hours. The last occurrence had happened in 1925, so it was anticipated that, if the legend were true, the seven would show up again on the night of the seventh full moon of 1932.

As it turned out, part of the legend was true. There are different versions concerning exact details by three living witnesses who were only boys at the time— Delta (my brother), Norman Littrell, and Lee Kennedy. I relate here what seems to be the most plausible composite of the three versions as well as my own recollection of what I remember being told. In any event, in mid-July, two days before the full moon, a group of Indians, all on pinto ponies, rode up to the cave's entrance and set up camp. Norman recalls there were eleven of them, all men; Delta remembers only eight, one being a woman.

The Littrells were using the mouth of the cave as a shelter for some of their dairy herd, and the Indian Chief complained to Ralph that the cave belonged to them. They wanted the cows removed and the cave fenced off to preclude entry of the cattle. Ralph did see that the cows were prevented from entering by erecting a temporary fence. He also provided them with some eggs, fresh beef, and firewood.

He attempted to find out what they were about, but could elicit no specifics, except that they would enter the cave on the night of the full moon for a ceremony. At no time during the next two days did anybody see any one of them enter the cave. Ralph asked if he might accompany them to view the ceremony, to which the Chief nodded in the affirmative and said, "But no come out." Ralph declined. When he asked when they would enter, he was told around midnight.

With Ralph driving the truck, Norman and Delta delivered milk early each morning. Having learned about the Indians' arrival, Delta asked if he could spend a couple of nights with Norman so as to be in on the excitement. He returned to the dairy with them after milk deliveries on the day of the full moon.

About dusk that evening, Ralph had his parents close down the dairy, douse the lights, and walk back across the river to their home on the other side. He spent the night in the hay loft so that he could observe the Indians from close range. Norman, Delta, and Lee Kennedy spent the night in the milk truck, parked where they could also view the activities. However, all but Ralph fell asleep and missed the action. He reported that an hour or so before midnight, with two Indians standing guard outside the entrance, the remainder, completely in the nude, entered the cave. Each carried a torch in his right hand and nothing in the left. At one point, one of the guards investigating the loft area almost found Ralph.

Several hours later, all who had entered came back out, dressed, attempted to block the cave entrance with some planking, cautioned some of those around the dairy not to enter the cave, struck their camp site, cleaned it up, and rode away before noon.

That afternoon the three boys carried planking into the mouth of the cave to make a walkway across the heavy deposits of relatively fresh cow manure. After the dairy chores were completed, Littrell, his sons, Ralph and Norman and son-in-law Herb Pritz, Oren and Lee Kennedy, and Delta entered the cave to investigate They found that in the center of a large room in the cave, a large fire had been built and was still smoldering. By the footprints in the dust, it was obvious that some kind of Indian dance had been performed around the fire. There were a few pieces of feathers and some small pieces of painted wood scattered about.

About 50 yards from the opening, the cave splits into two branches. The one to the left is the longest and has the highest ceiling. There were a few footprints indicating some activity had occurred in that area. The branch to the right narrows down quite abruptly into a smaller tunnel about 15 inches in diameter as far as could be observed with a flashlight. It appeared from footprints that it was through this narrow opening one or more had slithered, perhaps to recover from a secret hiding place some ceremonial trappings from which the feathers and small wooden pieces had fallen.

A discussion followed as to which one of the boys would also slither through it to see what could be found. Norman volunteered, but since he was rather on the heavy side, Delta, the smallest, was accorded the honor. A rope was tied to one foot and in he went. In about 50 feet the tunnel made a 90 degree down-turn, dropped about two feet, and ended. At the bottom, he discovered a down-covered nest (perhaps down from cottonwood trees) in which lay a cache of numerous and varied Indian artifacts, such as eagle feathers tied in bunches, beaded trinkets, colored wool yarn, and a number of unidentifiable carved and painted wooden objects. He gathered an armful and called to be pulled back out. Two additional trips were needed to recover all the artifacts.

Today, Delta regrets having disturbed this important Indian ceremonial treasure. At the time, he was only 13 and not mindful of what he was doing. Indians in those days were held in little regard and evoked only curiosity, not compassion and concern for their feelings. No one remembers with certainty what happened to the artifacts, but it is my recollection that they were donated to a museum in Santa Fe.

The Indians had come, it turned out, from Jemez Pueblo, about fifty miles west of Santa Fe, where the last survivors of Pecos Pueblo had moved when they abandoned their ancestral home some 91 years earlier. This same cave is probably the one mentioned in Willa Cather's book, *Death Comes for the Archbishop.*

A mere handful of the then Tererro residents know that in 1939, the Indians did return again (the mine had closed and only about two dozen people still remained in the area, Joe and I being two of them). We saw them encamped at the mouth of the cave, but made no attempt to spy on their activities. Three days later, they had departed. In all probability, they were angry and disappointed to find their ceremonial paraphernalia missing. Whether or not they returned in 1946, and thereafter is not known.

Delta worked on the milk truck for several years, and it was the only permanent job he had as a boy. I, on the other hand had many. My mother seemed to have the notion that I should be the town's local boy-entrepreneur, so she was always making arrangements for me to take on some new enterprise. The first was to sell the Sunday edition of the *San Francisco Examiner.* For three months, I would receive 20 copies each week which I was expected to go out and sell. They arrived with the comic section on the outside; the Katzenjammer Kids was the lead comic, although I believe it had a different title. I had about eight steady customers, could sell maybe another five copies, and was usually left with the remainder unsold. The enterprise was a financial flop, and she found something new for me to do after three months.

The next job was to sell Cloverine Salve which came in a flat, round tin. It was supposed to be a cure-all for chapped hands, of which there were many in the town, what with all the snow and cold weather. I think it was a fancy name for a Vaseline-like product, enhanced with the essence of sweet smelling clover, and I did a pretty good job selling it. It wasn't a good sales item during warmer weather, however, and she found three other items for me to vend: *Colliers, Liberty*, and *Literary Digest* magazines. Tererro was not a town with many people into the likes of *Liberty* and the *Literary Digest*, but *Colliers* did quite well. Unfortunately, I had to sell all three or none according to the agent who talked to Mom, so I gave up all three.

Next came the *Denver Post* which also sold well on Sundays, but few could afford a full subscription, and she agreed it was a losing proposition. We finally settled on the *Santa Fe New Mexican* as my youthful career in business, and I sold it until I went away to high school. Eventually, the load became so heavy that I had to give up a part of the town to another boy, Harvey Hurley. He, as I had, made enough to buy a bicycle, and then there were two of us who were the envy of the other kids.

One of the customers I transferred to Harvey failed to get his paper one evening, and thinking that I was still his carrier, he complained to Dad the next day at work. Dad came home, and without fully explaining the circumstances, gave me a razor strop licking for failing to deliver my papers properly. When I finally found out who hadn't received the paper, I was able to explain that the individual was no longer my customer. Dad didn't say he was sorry! When Harvey and I both went away to high school in 1935, both routes were taken over by Ralph Littrell, Jr. who also sold numerous magazines and the San Francisco Examiner. He was more ambitious than I had been, and he, too, purchased a bicycle.

My parents insisted that Delta and I invest the most of our earnings, before 1929, in a savings institution. They selected Liberty Mutual Savings in Denver. We each had accumulated well over $100, a sizable sum for eight- and ten-year old kids to have in those days, when Liberty Mutual and scores of other businesses, were wiped out in the stock market crash of 1929. Thereafter, our savings were invested in Postal Savings Certificates of $25 each.

chapter 22

* *

Youthful Escapades—Youthful Misdeeds

Joe and I seemed to have a penchant for getting into trouble. The first time, we were eight years old and had gone with my parents on a picnic near the Pecos River bridge close to Dalton Creek. As was often the case, Dad wanted to fish and Mom had prepared a picnic so that the family could enjoy the outing with him. Joe and I tried our hand at fishing too, but soon became bored and looked for something else to do. We spotted the Penitente cross at the top of the mountain, and decided to climb up to investigate. The ascent was approximately 2,500 feet, and required about two hours of climbing. On the way back down, I recalled how, on that first day we moved to Tererro, the boulder blocking the road had tumbled down the mountainside, flattening bushes and small trees in its relentless journey to the river below. I suggested it would be fun to see if we could find a few good size boulders to repeat the process.

After sending a half-dozen or so tumbling down the steep mountainside, we heard Dad shouting at us to stop immediately. He had climbed to a point about 100 feet below and far to the side of us so as to be out of harm's way. He was extremely angry and gave us a real tongue-lashing for our stupidity. One of the boulders had barely missed another fisherman who had beat a hasty retreat to safer ground. Also, Dad had had to give up the fishing he was so fond of.

In 1930, "Chief" Gonzales, husband of Matson's maid and cook, Jessie, purchased a brand new, lemon-yellow, Chevrolet coupe. It was equipped with what we then called balloon tires, and "Chief" would drive it proudly to the downtown area. It was garaged in what had once been the tack room at the ranch.

One hot afternoon that summer, Joe and I were looking at it, and for no reason that either of us could ever explain, we decided to deflate his balloon tires. Having done that, we decided the leather upholstery could use a bit of the car polish he had sitting on a shelf, and we liberally applied it. Finally, we put some dirt in his gas tank, then calmly went about doing something else to amuse ourselves.

That night when "Chief" came from work and discovered the blatant vandalism to his pride and joy, he went to tell Mr. Matson what he had found. Not suspecting that Joe and I could have been involved, he asked Joe if we had seen anybody around who might have done it. Joe promptly replied, "Leon and I did it." Within minutes, Mr. Matson, "Chief", and Joe were at our house, and the story was related to my parents. Try as they might, they could not elicit from us a reason as to why we would do such a thing. We simply did not have an answer.

Both of us were corporally punished—Joe with a willow switch and me with Dad's razor strop. In addition, we had to hand-pump the balloon tires while the three adults stood over us. Then we scrubbed and cleaned away the polish left on the upholstery, and since we could do nothing about the dirt in the gas tank, we had to wash and polish the car for "Chief" every week for two months. He had to have the gas tank removed to rid it of the dirt.

Our parents surmised that we had acted as we did because "Chief" had taken the place of our good friend Milton Henson; perhaps a psychiatrist today would concur, but in all likelihood, it was just something we did to amuse ourselves.

When we were ten, we engaged in another ill-conceived deed, but with much better end consequences. Near the golf course, which was uphill from the Matsons', Littrell grew oats for his dairy herd on a five-acre field. He had a horse-drawn planter at the field; it was left there all year rather than hauled back and forth to the dairy over two miles away.

Joe and I went to the golf course one Saturday in early spring to see if we could get a caddying job. Finding none, we walked by the oat field and observed this piece of machinery sitting idle and of no apparent value to anybody, since it never seemed to be used. We were intrigued by the wheels and gears, and thought perhaps we could use them to build something. We returned home to get wrenches and screwdrivers, then returned to the field and began dismantling the planter. When we had removed all the wheels and gears we could, we set out for home with our plunder, taking the shortest route to our house down the hillside, about a quarter of a mile away. The route led us along the rim of the little valley that broadened into the meadow upon which our house was located.

As we moved toward home, our load seemed to grow heavier and heavier, so we would toss one or two of the pieces down the steep hillside into the valley. It was covered with pine, fir, and aspen trees, as well as a dense undergrowth of weeds and scrub oak. By the time we reached home, we had abandoned our project and had discarded, one by one, all of the wheels and gears into the undergrowth. Surely, they could never again be found!

A few days later, Littrell arrived to plant his spring oat crop, only to find the planter in shambles. He was riding horseback and decided to ride down by the Matsons' to tell him what had happened. Joe and I happened to walk up about that time. Mr. Matson, as usual, inquired of Joe whether or not we had seen anybody dismantling the planter. Joe's customary and always truthful reply—"Leon and I did it."

For this we both received the usual corporal punishment with the willow switch and the razor strop. In addition, we had to search out and retrieve every piece we had tossed away—it took about a week—then under Littrell's supervision, reassemble the planter. He planted his crop about a week late, but said at the end of the season it was the most evenly planted crop he had ever had and credited it to our great mechanical skill.

At age 11 we concluded that we needed a clubhouse and workshop. We decided to purloin the necessary lumber from the mine's lumber pile, located on the south side of the mine complex. It was easily accessible from the trail leading to our houses. We chose a site on the top of a hill, somewhat hidden and out of view of our house, on which to do our building. The needed lumber was midnight-requisitioned and the clubhouse/workshop constructed in about a week. Jessie Gonzales' nephew, Pat Montoya, was spending the summer with her, and he helped us with our requisitioning and construction. We had an ample supply of tools, for Dad's toolbox was located at home; he no longer needed them on the job.

Getting electricity to the site proved to be somewhat more difficult. We located an abandoned power line near the water tank, shinnied up the poles, and pilfered as much as we needed to run a power line from our tent house to the site. The line was supported on pine trees at three locations, but because the trees swayed in the wind, we left plenty of slack to prevent the wires from breaking as they moved back and forth. I don't recall where we got the pole insulators, the light fixture, or the electrical outlets, but we did and we were in business.

How we ever thought our parents would not know what we had done is beyond comprehension. Fortunately, as we found out later, they talked over our activities and decided not to interfere. They reasoned it was a good idea to keep us occupied and out of mischief the rest of the summer, although we did get good lectures about taking things without permission. Mr. Powell, the Chief Electrician, came by to properly install our wiring because he said it looked like draperies hanging from the pine trees. Joe received a wood lathe for his birthday, and Dad helped us build a table saw by providing us with a saw blade and shaft, then showing us how to pour Babbitt bearings. We were kept busy building things not only that summer, but for several summers to come. Both of us became good craftsmen and, in later life, we both put our skills (not the stealing part) to good use in our own homes.

The only trouble we got into that was related to the workshop occurred when we were in the fifth grade. May Lynch (later married to Harold Walter) was our teacher. She was having us do a project on birds and suggested the boys might want to build birdhouses. She offered some small award for the best one. Richard Wheelock, whose father had just been employed as the mine Safety Engineer, won the award.

Joe and I thought it was unfair, for we knew that Richard's father had had my dad cut all the pieces for his birdhouse in his shop at the mine. We were new at this business of using power tools, and our birdhouses were quite inferior to his. We waylaid Richard on the way home from school and smashed his fine birdhouse on some rocks. Word got back to our parents, and again we received the usual corporal punishment. We also had to apologize to Richard. Unfortunately, he could have been a good friend, but for good reason he never liked us very well after that.

In 1937, when we both had learned to drive, Mr. Matson took us along with him on a trip to Shafter, Texas, to help drive. We stayed at Wheelock's home; they had moved there upon his promotion to General Manager of that mine. Richard still recalled what we had done to his birdhouse, and would have nothing to do with us. We decided to drive to Ojinaga, Mexico, about 18 miles to the south and just across the Rio Grande, to have a Mexican meal. We did not tell his dad what we were doing.

The highway to Presidio, Texas, on the American side of the river was beautifully paved, and a nice concrete bridge spanned the river to the Mexican border at mid-point. From there, the bridge was a wooden affair with small logs as the decking, and an unpaved dirt road. We drove into and around the village plaza. There were armed Mexican soldiers everywhere and, wisely, we decided not to seek out a place to eat. We drove back toward Presidio, where we expected to eat. At the border station, we told the

Immigration Officer our names and what we were doing. After telling us we were foolish for having gone to Ojinaga, because there was some kind of uprising going on which had brought in the soldiers, he sent us on our way.

About 100 feet further along, there was another inspection station which we assumed was a second border station, and we drove right on by. Whistles began to blow and a siren began to wail. We stopped, backed up and were chastised for having driven by the agricultural inspection station. When the inspector was assured that we had nothing to hide, we were told to drive on. Yet another inspection station appeared, and we foolishly thought, "Surely there wouldn't be a third place to stop," so we blithely drove on by. This time a patrol car took out in pursuit and pulled us over. We learned that we had passed the narcotics inspection station. After our explanation as to who we were and what we had done, the officer correctly decided we were just a couple of dumb teenagers. However, he took our names and information about where we were staying in Shafter before assuring us there were no more inspection stations and sending us on our way

Mr. Matson, who had received a call from the border station, was waiting for us with a well-deserved lecture for taking out on such a foolish trip without telling anyone where we were going. He correctly pointed out that, had we gotten into trouble in Mexico, we might not have been heard from again, since relations between the two countries was not too good at that time.

Halloween night was not a time for trick or treating, it was all tricking—one might even say it was a night for youthful vandalism, mostly the upsetting of outhouses and smashing of eggs on peoples' cars. Despite our previous escapades of mischief, Joe and I always went out alone and never did anything except observe what others had done or were doing. Only on one occasion did we decide to join in the fun—Joe carried two eggs in his coat pocket with the full intent that we would egg somebody's car. It never happened, Joe got egged instead.

We happened on to the local peace officer, a Mr. Malcolm, who knew us both well, and he stopped us for a friendly chat. He wanted to know what we were up to and, upon being told we were just out to see what was going on, he wished us well, but not before noticing the eggs bulging in Joe's pocket. With a friendly but very firm pat on Joe's side where the pocket bulged, he chuckled, "Well, boys, have a good time." We spent the rest of the evening back at my tent-house washing away the smashed eggs from Joe's pocket, so that his mother would not know he had filched them from her refrigerator.

Most of the boys would go around in a group, led by Roy Lynch, overturning every outhouse they could find. One night they overturned our workshop/clubhouse, spilling nails, screws, and tools into a conglomerate mess. It took days for us to sort them out after our dads helped upright the structure On another occasion, the group turned over an outhouse, and one of the boys, I believe it was Carl Jenkins, leaned too far forward falling into the pit below. They hauled him out, carried him over to the Pecos River, and dunked him in, clothes and all. When he was relatively clean, they built a bonfire and dried him out. He probably gave up turning over outhouses in future years.

Some of the girls sometimes accompanied the boys; Mildred Ballentine was one of them. The gang was about to overturn the Swenk's outhouse when he heard them, and blasted away at the group with a 410 gauge shotgun. They had all seen him raise the gun to shoot, and had turned to run away. Mildred happened to be in the line of fire, but was fortunately so far away that she was not seriously injured. It did take Dr. Smith an hour or so to extract the buckshot imbedded just below the skin of her backside. That incident put a damper on future overturnings, and Swenk was almost discharged for the incident. However, Tucker was then married to his daughter, and intervened on his behalf. He was warned never to use his shotgun on Halloween again.

One summer, several of us got interested in bows and arrows. We spent hours making the bows, once again from the Sitka spruce that dad furnished, and arrows from pieces of seasoned ponderosa pine. The tips of the arrows were not fitted with special arrowheads, but simply pointed. The flight stabilizers were made from chicken feathers. We used them for target practice and occasionally in an attempt to shoot a rabbit, chipmunk, or squirrel. One evening Delta and I were down in the garden and he was attempting to shoot arrows completely through a stand of corn. Most of them ended up stuck in a cornstalk, and it became an effort to find them. To save time, I volunteered to stand along side the stand of corn so as to be able to observe the arrows as they were shot through it.

The scheme was working well until one arrow ricocheted off one of the stalks and struck me in the corner of my right eye. Another late evening trip to the hospital. There was a time when the doctor thought I might lose the eye, but it eventually healed. Bows and arrows in our family immediately became forbidden.

Harvey Hurley was becoming quite proficient in archery and purchased a professionally made archery set with steel arrowheads. His archery days also came to an abrupt end when he shot an arrow at one of Littrells' cows. The arrow stuck in the cow's hip and Harvey took off in hot pursuit to retrieve it. The cow kept dodging away so that he could not grasp the arrow and, in frustration, he smacked the cow on the rump with the bow which broke in half. In disgust, he threw the remaining arrows away and gave up the sport.

Joe was not involved in the final misdeed of my youth, although he was, in an indirect way, somewhat responsible for it—he was in school in Roswell, I was in Santa Fe and, since I had no close friends, I missed his companionship. It was my junior year in high school when I bached with my brother and uncle. I did not enjoy going to school. One afternoon, I ditched my afternoon classes to go to a movie with an acquaintance, intending for him to write an excuse to which he would forge my uncle's name. We forgot to have him do so when we went to our separate homes, so I ditched the following day, and for the full six weeks that followed. I simply stayed in our room after my brother and uncle left each morning. Finally, I searched out the acquaintance, and had him write a note saying I had been sick with the flu. As planned, he forged my uncle's name.

After a couple of hours back in school, I was called to the office. Mr. Greiner, the principal, said he was quite concerned about my health and extended absence, and he wanted to discuss it with my uncle; he asked me where he could be reached. Reluctantly, I gave him the telephone number, dreading what was about to happen. He left the front counter to go make his call, and left the note lying there. I decided it was evidence that needed to be disposed of quickly; the best place was my stomach, so I ate it. Greiner returned saying that my uncle was out of his place of business, and would have to be called later. He started looking in his pockets for the note, and not finding it, asked me if I had seen it. Naturally I hadn't.

He sent me back to class, but called me back later to tell me the jig was up—my uncle knew nothing of my absence. He told me he couldn't understand why he wanted to bother with me, since I would never amount to anything anyway, but he meted out my punishment—I would have to attend eighth period (the normal school day was seven periods) for the remainder of the school year, another six months.

He was a wise and astute principal—his words challenged me to show him he was wrong. I made all F's for the grading period during which I had been absent, but in an attempt to show Greiner I could amount to something, I made only A's and B's during the remainder of my high school education, and had a perfect attendance record during my senior year.

Twenty-eight years later, I saw an article in our local newspaper, including a photograph, stating that Bright E. Greiner was retiring as principal of the local adult education department of our school system in Fairfield/Suisun, California. I immediately recognized that he had been my high school principal and gave him a call. I told him my name and averred that he probably wouldn't remember me. He immediately replied, "Oh yes I do remember you. You were the boy who ditched school for six weeks straight." We had lunch together a few days later, and had a good laugh about the incident, especially when I told him I had eaten the note.

One day before we became teenagers, Joe had accidentally started a small brush fire near his house. From that time forward, if there was ever a brush or forest fire in our area, he and I were considered prime suspects as youthful firebugs. Although we never started any, we were sometimes the first to arrive on the scene. In late June, 1938, we were enjoying a malt at Picks' soda fountain when someone came in and said there was smoke coming from a small side canyon off from Louis Rivera's property. We borrowed a rake and shovel from the store and took off to see if we could help. On the way, Rivera's son, about age 13, joined us, and we were the first three to arrive at the fire around three p.m.

Within a half hour, Mr. Johnson, the Forest Ranger, arrived with a crew of recruited men and took charge of the operation. Fortunately, because we had arrived in time, we had prevented it from spreading as far as it might. By about eight p.m., it was contained and under control. Johnson decided the older men all had jobs to go to, so relieved them of duty, but kept Joe and me on the fire line to insure that it would not flare up again. Not many know that a Forest Ranger has that authority. Rivera's son was also relieved of duty, but was asked to go back with Johnson and return with some food for us; this he did around ten p.m. Johnson returned around seven a.m. the following morning, assured himself the fire was completely out, and let us go home. Our parents were frantic with worry when we returned, bone-tired and weary, with blistered hands. They had no idea where we had been. We were both fed, our blistered hands cared for, and put to bed.

It was the policy of the company that any employee's son who was a high school graduate would be given a job if he wanted one. Since I had graduated that spring, I was eligible; Joe had a year to go and not eligible, but we tried to convince Mr. Matson that both of us should be hired. Mrs. Matson was adamant that she didn't want Joe working and, because I was only 17, she thought neither of us were physically capable of working an eight-hour per day job, so neither of us were hired.

Late in the afternoon after the fire, Joe came running over to tell me his mother had relented, and we would be hired. She had decided that if we could be recruited to fight forest fires, we would be safer working at the mine. The following morning we went to see Ray Marsten, the Master Mechanic, and were employed as general laborers. We didn't know what we were getting into.

Since our hands were sore from the blisters, our first job was to go through the shops, collect all the scrap copper wire, and strip the insulation so that it could be sold for scrap. After the first week, we were assigned as helpers to the two men who collected the garbage on Monday, cut firewood on Tuesday and Wednesday, and did other odd jobs the remainder of the week, such as crush rock to make gravel for the roads, or mix concrete by hand. The garbage had to be incinerated beginning on Tuesday. We inherited that job—starting a wood fire in the incinerator each morning, then shoveling in the garbage all day, and cleaning out the non-combustibles in the late afternoon. If it rained in the afternoon—it usually did—the garbage would get wet and might require three days to completely burn. The maggots were always profuse and quite sickening, but we eventually got used to them.

Helping to cut wood was a boring, muscle-building, tiring job. The two men operated the cutoff saw while Joe and I had to feed them the four-foot-long split timbers that had been seasoned in stacks six feet high. Every eight feet in the length of the stack represented a cord and one-half, and we would cut about twenty cords during the day. The saw would be moved closer to the supply about every two hours, so there were times we were carrying the heavy lengths about 20 feet, at other times, only about 6 feet. To keep up with the men, we had to almost run at the times the supply point was distant from the saw; we were always glad when the supply point was just about six feet away.

Joe had to quit on September 2, to return to school and I worked until September 30, when I was laid off because the new wage-and-hour law became effective; anyone not yet 18 was barred from working in a hazardous industry.

He and I were two very lucky boys, for had Mr. Matson not been the mine manager and the ultimate authority in all that transpired in Tererro, we may very well have ended up in a reform school as a result of our transgressions. However, all of the transgressions, except the trip to Mexico, happened before we became Boy Scouts, so we can probably attribute our turn around to the good influence of the organization, and the excellent leadership of Dan Rogers and Charlie Barrett.

chapter 23

* *

Pleasant Diversions

Although living conditions in Tererro were not ideal, residents did have a number of outside interests to occupy their free time. Golfing was one of them. Early in 1929, a management decision was made that the company should provide some kind of recreational facilities for employees and their families. The initial project was the construction of a nine-hole golf course, that was completed and ready for play by mid-July. It had few of the amenities found on golf courses elsewhere, but in many ways it was more challenging than other courses.

Due to the short growing season and cold winters, the greens could not be covered with a verdant grass turf. They would only be green about three months out of the year—the remainder of the year, they would have been covered with dead thatch. Consequently, the "greens" were composed of oil soaked sand, the oil being the waste oil drained from engine crankcases.

The ratio of oil to sand was varied from green to green, as well as on different areas of each green, so as to create a diverse modulus of compaction. At each green, there was a drag made in the shape of a "T," the handle being a one-half inch steel pipe. This was welded at the mid-point of a 24-inch long, one and one-half-inch pipe. Each player would use the drag to smooth a pathway in the oil-soaked sand, dragging from the lay of the ball on the green to and beyond the cup. Caddies had to remove any sand from the cup which had been dragged in. The pathways, due to the varying density of compaction might be quite firm, and the ball would require a more gentle stroke of the putter. On the other hand, the pathway might be quite soft, in which event the ball would require a firmer stroke of the putter. In some cases, there would be both hard and soft compacted areas, creating an entirely different putting situation.

One might ask, "Why not simply putt without dragging the pathway?" The answer was quite simple—people, cows, dogs, deer, horses, or other animals frequented the course and walked across the greens leaving deep footprints in the sand, and there was no way a reasonable putt could be made without smoothing its way to the cup. The course was not fenced in and Littrell's dairy cows roamed and grazed the course daily—just one of the hazards of play. There were no such things as sand traps—fresh cowchips were hazard enough, especially when deposited on a green.

In that mountainous area, the terrain was not gentle and only four fairways, numbers 4, 6, 7, and 8, were somewhat level. Fairways 1 and 2 were steeply inclined uphill, and fairway 3 steeply declined downhill. Fairways 5 and 9 were normal to the slope of the mountainside, and a driven ball on number 5 could easily begin to roll downhill, crossing fairway 9, and coming to rest on fairway 6. Golfers with experience on regular turfed greens in Albuquerque and Santa Fe found the par 36 difficult to achieve.

Because bag carts and golf carts were yet to be invented, most players chose to hire a caddie to carry the bag of clubs and to spot balls. The going rate was 25¢ for nine holes, and that is where many of us boys earned spending money. I was age eight when I began, and would sometimes, as did other boys, caddie for two players at one time carrying two heavy bags of clubs—they would often play a total of 36 holes. The normal pay would be $2.00, but on occasion we would receive a tip of up to $1.00.

Most men worked six and seven days a week, so Sunday was generally the only day they could play. Some of the more avid golfers could find time for about six holes after work, and some office employees occasionally could play on a Saturday afternoon. About a dozen women took up the game and would play during the week; but, in the main, the course was free for kids to use Monday through Friday. None of us had more than four clubs—a driver, a five iron, a seven iron, and a putter, but we did become very proficient with those four, and most could finish a round in 38 to 45 when we reached our middle teens. Had we had the opportunity, we could well have beaten many of the men. However, when they played, we caddied.

In 1931, a baseball diamond was built on the crest of the hill adjacent to and west of the golf course. The infield was comparatively level, but the left outfield declined toward the golf course about four feet at its furthermost edge and the right field declined toward Rico Town by about two feet. A hard-hit ground ball into left field gained considerable momentum as a result of this slope. For a return throw to home plate of a ball to deep left, the fielder had to throw uphill about five feet.

Gilbert coached the Miners, the ball-playing employees of Amco. They competed in the Northern New Mexico League with teams from Albuquerque, Madrid, Santa Fe, Raton, and Las Vegas. The Miners won the championship three out of the seven years they played in the league, but the most thrilling championship game was played in 1934.

The Albuquerque Dons were tied with the Miners for first place going into the last game. At the end of the first half of the ninth, the score was tied five to five. At the bottom of the ninth, the first ball pitched to Joe Lucero, the Miner's pitcher, was a resounding home run far into right field, and the Miners had won their first championship.

It has been alleged and published in a couple of magazine articles that the company hired men, if they were good ball players, on the basis that they need not fear loss of their jobs even if they were incompetent as employees. It makes for a good story, but it simply was not true. Joe questioned his father about this when the story was first published, and he emphatically denied that it was the policy. He said that Amco was in a profit-making mining business, not the baseball business; every employee was expected to produce at an acceptable level—even ballplayers. He did indicate, however, that if there was a choice between a good worker only, and one who was good at both mining and playing ball, the ball player would have been selected. This seldom happened: good miners were simply too difficult to find, so both would have been hired.

In 1932, a clay-surfaced tennis court was built just north of the messhall, but there were few tennis players to use it. Dan Rogers tried to get some of the Boy Scouts interested without much success, and sometimes the court sat unused for weeks on end. The most I ever observed it being used was during the summer of 1939 after the mine had closed.

Les McClure's brother-in-law, who coached tennis in the Dallas school system and who was involved in operating the McClure Restaurant after the mine closed, would challenge anybody who came in the restaurant to a match, and he seemed to find no dearth of challengers among the "dudes" at Cowles and summer residents in Holy Ghost Canyon. Every day, early in the morning, about mid-afternoon, and later in the evening, he would be on the court with one challenger or another. He said that it was the first clay court on which he had ever played and quite a challenge to master.

During the winter, some adults turned to bridge and dinner parties for entertainment. The first few years, auction bridge was in vogue, but quickly gave way to contract bridge in 1932 with the publication of the book, *Contract Bridge Blue Book* by Ely Culbertson.

However, family entertainment, most evenings, centered around the radio, listening to *Bing Crosby, Amos 'n Andy, Bob Hope, Major Bowes Amateur Hour, Little Theater off Times Square, The Lucky Strike Hit Parade, The Shadow, One Man's Family,* and other programs of the era. The very first radio program I remember hearing was the Dempsey-Tunney heavyweight championship fight on the night of Sept. 23, 1926. The only radio in the new community at that time belonged to a friend of Dad's who lived in the same bunkhouse that he, Dad, had occupied before the family moved there. His friend invited Dad to come hear the fight, and he took Delta and me along. I was too young to know what the excitement was about, but I do remember well all of the men listening intently and cheering on their favorite, and at the end, arguing about a long count.

Radio reception could be received only on three stations: KOB, Albuquerque; KOA, Denver; and KSL, Salt Lake City. To have decent reception on any of the three, it was necessary to string an antenna wire, at least 100 feet in length, between two tall pine trees. Due to numerous lightning strikes during the many summer thunderstorms, it was also essential to have an antenna bypass switch from the radio to ground, and to make sure it was switched to ground when the radio was not in use, or whenever a thunderstorm was in progress. An antenna high in the tree tops was more apt to attract a strike than was a tree if the lightning was nearby, and many learned the hard way by having their radios blasted into hundreds of pieces when they failed to make the switch.

Fishing and hunting were probably the two outdoor activities in which the majority of the men, as well as a few women, participated. Deer, elk, and grouse were hunted by many, both in and out of season, to put extra meat on their tables. As a result, the deer and elk populations declined. The deer that had gathered at the "deerlick" each evening during the first three or four years eventually ceased to come to their centuries-old salt lick.

The decline was particularly noticeable with the elk. They had been reintroduced to the area in 1915 after being hunted to extinction about the time Case came upon the ore outcrop which eventually led to the founding of the town. Fifty head were shipped from Yellowstone National Park, but only ten bulls and 27 cows survived the trip to Grass Mountain where they were released to shift for themselves and procreate. Procreate they did; in 1932, the herd had increased to between 200 and 300, and were scattered throughout the wilderness and high country area. However, their numbers began to decline again in the Cow Creek, Willow Creek, and Bear Creek areas when the mine opened, areas which were easily accessible to a few men in Tererro who had little concern for preservation of the local natural resources. Even though there was no open hunting season for elk, they were illegally hunted during the winter months when the meat could be frozen outdoors and used before warmer months set in.

Only three or four families had a mechanical refrigerator; others had to get by with ice boxes, the ice being sold by the dairy. We had the good fortune of having the use of an ice storage house at the Matsons since, because they had a refrigerator, they didn't use it. We collected ice from the overflow of the water storage tank. Dad was not a hunter, however, so the ice was not used to preserve wild game.

In 1933, the State Game Commission decided the time had come when a limited number of permits could be issued for a ten day hunt of bull elks with at least three tines on each antler. One hundred permits were offered, but only 97 were issued; at the end of the ten days, only three elk had been taken, the first one by Ole Lee. He had the head mounted, but donated the meat for a community-wide barbecue.

Near the messhall, a large pit was dug and lined with river boulders. Logs were burned in it for a full day. Early that evening, the whole dressed carcass, about 600 pounds, was placed on a spit over the heated rocks and glowing embers, and slowly cooked for about 15 hours. In the messhall, meanwhile, salads, baked beans, and hot breads were prepared. At one p.m. serving began; by five p.m., the then approximately 2,000 residents, as well as invited guests from the surrounding area and the mill, had feasted on what, to most, was their first and last taste of elk meat. The event was courtesy of Amco and Ole Lee.

Black bear were also hunted when in season, but were probably not hunted illegally to any great degree. They were around, but not in great numbers. One afternoon when we still lived on the twin-bridges meadow, Mom took us children with her to gather wild gooseberries. The bushes were sometimes quite large and dense. As we picked on one side of one of these large gooseberry thickets, we heard what sounded like somebody else picking on the opposite side. Mom went around to investigate and almost ran into a bear raking the berries into his mouth. We retreated as quietly and as quickly as we could. None of the children saw the bear, but Mom was thoroughly startled by her near-encounter with it.

Most men fished during season, some probably out of season, in the Pecos, Holy Ghost Creek, Willow Creek, Rio Mora, Winsor Creek, and other tributaries of the Pecos which were easily accessible. Trout were abundant, and it was not too difficult to supplement food supplies with daily limits of 25. Dad fished several times each month, rushing home from work and heading for a favored fishing spot to bring home enough for our evening supper. Trout served in restaurants have never been as tasty as those caught in the cold, often icy, mountain streams.

About 1935, a company employee, Bob Nelson, converted an old barn owned by Christino Rivera (father of Louis Rivera), and located at the edge of his property along the Cowles road, into a movie theater. Movies were shown on Saturday and Sunday nights. In addition to a feature film, there was also a continuing serial such as Buck Rogers or Flash Gordon.

His wife also taught tap dancing to some of the young girls and a couple of boys. Late in August, she would have a dance recital in the theater on a Friday night. They had a beautiful and precocious daughter, Dixie, who at around five years of age was quite the hit of the recitals. When the mine closed, the Nelsons moved to California. He went to work in the aircraft industry and his wife, Maymie, managed to get Dixie a screen test. She eventually became one of Hollywood's young starlets under the name Lori Nelson, and was featured in several movies during the 1950's and 1960's.

Danny Dew, a messhall cook, was an amateur magician and, on three occasions, he entertained the community with his feats of magic. As with all magicians, he would involve people from the audience, particularly the kids, and we all clamored to be selected—I never was. To most of us, his most profound trick of all was the snapping together of numerous 10-inch diameter endless silver rings. We were all disappointed when he transferred to the mill as a cook, and carried on his feats of magic for mill residents.

As at Tererro, there was both a nine hole golf course and a baseball diamond for employee recreation at the mill. Since mill residents were also provided with a clubhouse, they could regularly enjoy large-group indoor activities such as square dancing, ballroom dancing, bridge, and particularly competing in bridge tournaments. There developed a lot of rivalry with bridge groups from Tererro, and eventually with some from Santa Fe and Albuquerque. Jimmie Russell and Ole Lee from the mine developed into an almost unbeatable team. Jimmie laughingly tells how the two of them developed a somewhat shady signaling system which kept them a tad better equipped to compete than their rivals. It was probably not as sophisticated as systems later developed by Goren and other bridge experts, but their rivals were at a distinct disadvantage; they didn't know the Russell-Lee bidding system.

The power plant spray pond at the mill was about five feet deep, and one end was void of any cooling sprayers. Those living at the mill were permitted to use that end as a swimming pool, and those of us living at the mine were occasionally fortunate enough to get to go to the mill for a swim party. That is where I learned to swim, and for several years, I had no fear of the water.

A favored meeting place for residents, teenagers as well as adults, from both the mill and the mine, was Wynn's Casanova in Pecos where one could dine (somewhat elegantly), dance to music from a Wurlitzer juke box, play slot and pinball machines, and after prohibition was abolished, imbibe in alcoholic beverages. It was, without question, the most urbane place to socialize and eat other than establishments in Santa Fe. However, even urbane might be considered too refined a description when considering the total rural setting in which it was located. Although many teenagers did drink beer (there was no attempt to control the practice), Joe and I were content with an occasional hamburger, a malted milk, some music on the Wurlitzer and squandering a few nickels in the slot or pinball machines.

The pump station on the Pecos, near Valley Ranch, was adjacent to a large, flat meadow that was the Burrell's front yard. Burrell was operator of the pumps from the time they were turned on in 1926 until they were turned off in 1939, and he and his wife lived in a house provided there by the company. The large, grassy meadow was room enough for large groups, and the Burrells hosted many fish frys every summer for both mine and mill residents. Catching fish in the Pecos was relatively easy in those days, and people would combine their catches of trout for the occasions. It provided one of the best opportunities for socializing between the people from the two communities. Often there would be a hundred or more in attendance. Since most couples had three or more children, well over half of those in attendance were under age 14 or 15. The older teenagers usually had something more important to do.

Boy Scout Troop 21 was chartered and sponsored by Amco in 1931. Joe and I were still too young to join, being only ten-years old at the time.[1] One of the office employees, Dan Rogers, was paid a little extra to be the scoutmaster, and all expenses over and above the cost of each boy's uniform were borne by the company. Each boy was encouraged to earn a little to pay a small monthly dues. Failure to do so was not cause for being kept out of the troop, however. Every boy over 12 was encouraged to participate; unfortunately, very few of the Hispanics felt comfortable enough to join, due to the animosity between the Anglo and Hispanic boys.

Each year, shortly after school summer vacation began, the troop's expenses were paid by the company to participate in the annual Boy Scout Council camp held either in Tesuque Canyon east of Santa Fe, or at Los Alamos Boys School northwest of Santa Fe[2] , or in some other camping experience. The troop ranged in size from 15 to 20 boys. Transportation to and from these events was in one of the company's stake bed trucks.

In 1932, a trip to the Carlsbad Caverns was arranged. Several of us who were not yet old enough to join were permitted to go, making a total of 23 boys; two trucks were required for the trip. Dan Rogers drove one, and Charlie Barrett, another office employee, drove the other. Charlie became the assistant scoutmaster, and eventually, the scoutmaster. The trip, one of the highlights of our scouting experiences, took five days. We camped out in our pup tents and cooked all of our meals except the one we ate in the restaurant within the caverns.

In 1933, the first year Joe and I were Scout members, the troop was taken on a tour of the mine that lasted about three hours. About 18 boys and our two leaders were outfitted with hard hats to which carbide lamps were attached. We were divided into groups of ten. Each group of ten was then loaded into a cage and transported to the 1,200 level. Tucker was the tour guide for one group, and Gilbert was the guide for the other group.

At the 1,200 level we were guided through various drifts to observe the pumps that helped keep the mine relatively dry, and to bodies of ore being stoped. There we observed miners blockholing by drilling for the emplacement of dynamite, preparing the dynamite by inserting the blasting caps attached to long lengths of fuse, emplacing the dynamite into the previously drilled holes, and stemming them with sand. We were then taken to a safe area while the blasting round was detonated.

We were shown how the miners checked the broken ore to insure it would be segregated from any gangue. We watched the muckers loading the ore into side-dump cars for transportation to the ore pocket See Fig. 11 at the main shaft station. At that station, we saw how the ore was loaded into the skips for transport to the surface. Later, we

---

[1] At that time, cub and explorer scouts were not a part of the scouting program, and 12 was the entrance age rather than 11 as it is today.

[2] Robert J. Oppenheimer attended these camps at Los Alamos when he was a Boy Scout, and because he knew how inaccessible it was in those days, he recommended that it be selected for the site of the atomic bomb project during WW II.

would be shown how the skip would automatically unload into the crusher ore-bin at the surface.

The diamond drilling operation was fully explained to us by Mr. McCloud, a Canadian under contract by the company to conduct all their diamond drilling exploration. He explained how the diamond bits were made and how the cores were identified and placed into core trays when the cores were removed from the bit.

Although most of the tour took place at the 1,200 level, we were shown other mining operations at the 800 level also. There we saw how timber sets were emplaced both in drifts and in stopes for raises. Wooden wedges were used profusely to raise vertical timbers so that the tops of all were in the same plane, and to insure snug fits.[3] We also saw how stopes were gobbed after all ore within the stope had been removed.

To reach the 800 level from the 1,200 level, and also from the 800 level to the surface, we were not transported in the cage—we climbed up manways on the ten-foot long wooden ladders my father had made. At every ten-foot level, there was a landing cut out of the rock where one could rest before ascending the next ladder above. The total vertical climb would be the equivalent of climbing to the top of a 120-story building on ladders attached to the face of the building, and resting on each window sill. Although we were all used to climbing mountains of greater height, none were ever so difficult, because in mountain climbing each step might be a small rise and the lifting of one's body only a few inches, whereas on the ladders each step was twelve inches higher and one's body had to be lifted by that amount.

At the surface, we viewed the mechanism that automatically dumped the skip, then toured the crushing process from beginning to end, and the tramway loading and launching operation. The tour ended in the hoist room where the operation of the hoists was explained, particularly as to how the lifting of one skip and cage was partially counter-balanced by the simultaneous lowering of the adjacent skip and cage. We were a tired group of boys, and were taken to the change house where we could shower and change to clean clothes. Working in a mine, we had discovered, was an extremely dirty business, particularly with silt-laden water dripping on you everywhere you went.

Two years later, we went to a camp in Tesuque Canyon on June 7. I recall the date clearly because it was Joe's 13th birthday. The weather was warm, and we all wore our light-weight, short-pant uniforms which were mandatory for that kind of weather. The pup tents were only long enough to cover our upper bodies, and therefore, the lower part of our sleeping rolls (sleeping bags had not yet been invented) were exposed to the open. On the following morning, we awoke to find that the exposed part of us was covered with about seven inches of new fallen snow. The temperature had dropped to around 25 degrees Fahrenheit. None of us, scoutmaster included, was equipped with clothing to withstand that kind of weather. Because our transportation truck had returned to the mine, there was no way for us to go home. The other troops all had transportation available in Santa Fe, and were immediately evacuated.

---

3 When the major construction at the mine was completed, my father's work as overall construction supervisor was completed, and he chose to stay on at the mine in a lesser position rather than seek a new job. He was in charge of the carpenter shop and personally responsible for making all wooden ladders, thousands upon thousands of wooden wedges, and replacing the skip and cage guides when they had worn beyond safe operating limits.

With blankets wrapped around us, we shivered around campfires until about noon when someone from Santa Fe arrived with a truck to transport us to the gymnasium at St. Michael's School for Boys in town. Snowfall had been so heavy in the Pecos Canyon that it was impossible for the company truck to come for us until the following day; we had to spend the night in the gymnasium. There being no place to cook, we had only cold sandwiches for food. That experience, unlike the one to Carlsbad, was one we all would have been happy to have missed.

Living as we did within a National Forest and with the Pecos Wilderness[4] only six miles distant, we could not have had a better setting for a Boy Scout troop. Whether we were on a day, or an overnight hike, the scenery and the setting was always magnificent. We had unlimited opportunities to earn our merit badges in woodsmanship, cooking, camping, ornithology, flora and fauna, first aid, camping, forestry, etc. Most of us advanced to "Life" rank as rapidly as permitted in the scouting by-laws, but getting to the "Eagle" rank proved almost impossible, since we all had to leave home at age 15 or 16 to attend school after completing the 10th grade. We would have had to join a new troop where we attended school, and where the opportunities for getting merit badges was more limited. New Mexico Military Institute had its own troop which Joe joined, and he was the only one to reach the rank of "Eagle." He also attended the National Jamborees in Washington D. C., in 1937 and, as a scoutmaster, at Valley Forge, Pennsylvania in 1950.

Because of the camping expertise we had learned from Milton Hensen, our parents began allowing Joe and me to go on nearby, overnight campouts by ourselves when we were ten. Each succeeding year, we could go further afield, and because we did well in earning camping merit badges when we became scouts, at age 13 we were permitted to take our first trip alone into the Pecos Wilderness.

He had received a horse, Charlie, for his seventh birthday that was stabled in the barn and pasture behind the Matson house, along with two company-owned horses, Blackie and Nebbie. On that first trip into the wilderness, we rode Charlie and Nebbie to Stewart Lake where we fished and camped overnight. The following summer we were out three times and ventured as far as Lake Katherine on the last trip. Everything went so well that in June of 1936, we planned a five day trip to Lakes Katherine, Truchas Lake, and Pecos Baldy Lake.

Having three horses available, we invited a friend, Wilmer Easley, along who rode Blackie. We arrived at Lake Katherine early in the afternoon, removed the bridles and saddles, then hobbled and tethered the horses to a log. We ate lunch and went fishing. About five o'clock, one of us went to move the horses to a new grazing location only to find all three missing, as well as the log to which they had been tethered. Never before had this been a problem with Charlie and Nebbie—they always stayed nearby. Apparently Blackie had decided to go home, and since all three were tethered to the same

---

[4] Although the National Wilderness Preservation Act was not enacted until September 3, 1964, at which time the name Pecos Wilderness, as we know it today, became the official name within the context of the act, in the early 1930's, 133,000 acres had been designated the Pecos Wilderness Area by the U. S. Forest Service. This was the third of three such areas so designated by the U. S. Forest Service, the other two being in Colorado and Montana. Basically, the same rules applied to these three areas as those enacted by the Congress in 1964. The size of the Pecos Wilderness was almost doubled with passage of the act.

log, the other two followed along, dragging the log behind them. We rushed off in hot pursuit hoping to overtake them before they had gone too far.

The last quarter-mile of the trail to Katherine is extremely steep. It weaves in and around many large boulders and we were quite sure the three horses tethered together could not have negotiated the trail without the log catching on a boulder or some trees. We were not to be so lucky. We found the log about a mile down the trail from the lake and gave up the pursuit. Three very dejected boys returned to the lake and began making plans for the next day. Early next morning, we secreted all our gear in an out-of-the-way spot, buried the fish we had caught in a snow bank, and began the long hike home. Our shoes were not suitable for hiking; we ended up with severely blistered feet. The horses had not returned home, as we had hoped they would, when we arrived.

The following week, the two of us hiked the six miles to Cowles, rented three horses from Viles' Ranch, and rode bareback to Katherine to retrieve our gear. One does not ride a horse bareback up the last half mile to Katherine—it is too steep to stay on the horse, so we walked and led the horses. Mr. Matson sent a truck from the mine to meet us when we returned to Cowles late in the evening.

Each winter, the three horses were wintered-over at the Martin ranch on Cow Creek, and it was to there that they headed. About three weeks later Martin called Matson to report he had found them grazing with his horses, their hobbles broken, and the tether ropes still around their necks. Again we had a long hike ahead of us—almost ten miles over the mountain to the Martin ranch, carrying bridles with us. It was a long hike of about six hours. Mrs. Martin gave us a good supper and bedded us down in a spare bedroom. The next day we returned home riding bareback again. We had had enough of horses that year.

Shorty Gallegos, knowledgeable in the ways of horses, advised us that in the future we should take along a bag of oats and give each about a quart upon arrival at a campsite. He assured us that they would not stray far from camp knowing that oats were close by. It worked!

On May 29, 1937, we set out on our biggest adventure of all. Using Blackie as a pack horse (the pack included a bag of oats) we planned a six-week trip into the wilderness, one in which we planned to camp at all the lakes (Katherine, Stewart, Johnson, Pecos Baldy, Truchas, and Lost Bear), and to climb all the peaks (Santa Fe Baldy, Capulin, both Pecos Baldys, all three Truchas, Chimayosos, Jicarillo, and Barbara). We carried food enough for one meal each day and planned on catching trout for the remainder. If we failed to catch enough trout and ran out of food, we would simply cut the trip short and come home. Although we did end the trip in about five weeks, it was not for lack of enough fish—we had trout almost every breakfast and for every supper. We gave up because rain prevented us from climbing Capulin and Santa Fe Baldy. In addition, we had already been to Santa Fe Baldy's summit on an earlier trip to Katherine, so missing that climb was not a disappointment. The final, and probably the overriding reason, was that we decided we wanted to be home for the Fourth of July and the fireworks display the Matsons always had; the McDuff family was always invited to come watch.

At the top of South Truchas, we carved our initials in a post held up by the cairn at its highest point. In doing so, Joe cut his finger, not seriously, but the bleeding simply would not stop—there was too little oxygen for the blood to coagulate. I tied a tourniquet around his arm and we decended as quickly as possible to our campsite at Truchas Lake. The bleeding had stopped by the time we arrived.

At Mr. Matson's suggestion, we were always on the lookout for outcroppings of ore as we climbed around on the mountain sides. On the way to the top of Pecos Baldy, I happened to pick up a piece of quartz as we climbed, not because it might be ore bearing, but because it was pretty. Only after we stopped to rest did I examine it closely. To my surprise, there was a small gold nugget in it, about the size of a pea. We marked the spot where we rested with a rock cairn, but did not return to look for the locale where I picked up the quartz.

Upon returning home, I gave the quartz to Mr. Matson who in turn had it assayed. Several other small specks of gold were found within its body, and it assayed at a high level of gold—I can't remember the exact figures. Early on the morning of July 12, Chief Engineer Anderson accompanied Joe and me back to Pecos Baldy, and we began a search for the resting spot we had previously marked. Our original route-of-climb had meandered back and forth across the face of the mountain, so it was difficult to know just where we had been. Although we thought we had carefully picked some landmarks to lead us back to the marked spot, we could never find it. About mid-afternoon, snow began falling, and Anderson decided to abandon the search. By the time we reached the horses at the lake, there was at least two inches of snow on the ground.

Almost certainly the quartz piece I picked up had been dislodged from an outcropping higher up the mountain side, and had we located the spot where I picked it up, we might still have not been able to find that outcrop. In any event, there is no doubt that somewhere on Pecos Baldy such an outcrop of gold bearing ore does exist. It will remain there forever unless, by some fluke, another stumbles upon it. One of my uncles tried finding it for years without luck.

While camping, horseback riding, fishing, and an occasional swim were the primary recreational activities in the summer, ice skating, skiing, tobogganing, and sleighing were our principal winter activities. The best place close to home for skiing, sleigh riding, or tobogganing was between our house and the Tucker's. Kids from all over town would come on weekends or moonlit nights to sleigh. One could begin at the road passing in front of our houses and ride down the hill to the lower end of the garden. This was usually the end of a fast ride of about 400 yards. When conditions were just right, one could go even further—all the way to the mine on the pathway leading there—about another 400 yards.

At the mine, the usual end of ride was on the trail just above the warehouse; few ever turned down a sharp incline onto the end of the mine roadway. One day, however, riding on my belly, I turned down the incline just as Billy Montgomery drove around the end of the warehouse, planning to stop, back-up, and turn back to the front of the warehouse. My route of travel placed me directly toward the center of his car, and there was no time to veer to either side. Fortunately, the underside of his car was high off the ground, and I had just enough room to pass completely underneath the car unscathed. We were both severely frightened by the incident, he so much so that he couldn't speak. I quickly retreated toward home carrying my sled (a Flexible Flyer) under my arm knowing full well that I was in for an unpleasant discussion with Dad that evening. It wasn't as bad as I had anticipated, but the next day a barrier had been placed to prevent further such occurrences.

The only disagreement I ever remember having with Joe occurred on that mine pathway after we had ridden our sleds down it one day. On the way back, we had a friendly scuffle in which he pulled off my boot and threw it down the hill below the path. Snow was several inches deep, and I would have had to retrieve the boot by climbing down the hill with only a sock on one foot. I tried in vain to get him to retrieve the boot and, when he wouldn't, became angry with him. The friendly scuffle became more serious and we exchanged a few blows. Finally I told him I was going home without my boot, would tell Mom what happened, she would tell his mother and he would be in trouble. He vowed that I wouldn't because my foot would be too cold from walking in the snow. To show him that I could, I placed my stockinged foot on my sled and began moving it along by pushing with my other booted foot. After several yards, he decided I was serious, and returned to get my boot. We made up and never again had another disagreement.

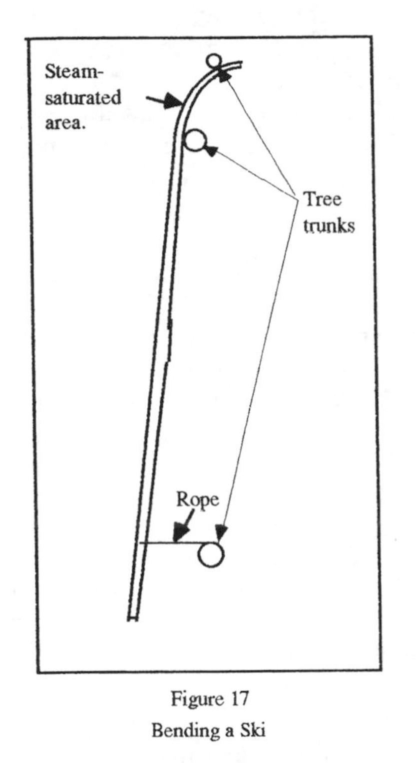

Figure 17
Bending a Ski

Very few of us owned skis, (store-bought ones were far too expensive), so those of us who did have a pair had made them ourselves. Dad, who as previously stated, was responsible for maintenance of the cage and skip guides in the mine shaft, would occasionally salvage some of the guides when they had worn beyond use. He would rip them into one-inch thick by four-inch wide boards from which we would make the skis. The Sitka cypress guides were ideal for this purpose—the wood was strong and flexible, there were no knots, it had been kiln dried, and it was very close grained. Shaping the boards into skis was a long and tedious job. From the center where the feet would rest, the boards had to be taper-planed to a half-inch thickness at either end, and the front ends rounded to a point. The leading ends were then wrapped in wet gunny sacks, and steamed over a tub of boiling water for at least six hours. To keep the steam rising, the water had to be in a rolling boil, so the wood fire under the tub had to be constantly tended. Water at that altitude boiled at considerably below 212 degrees Fahrenheit, therefore the steaming process took much longer than it would have required at sea level.

When the ends seemed to be fully steam-saturated, the tips were bent by placing the tip end against one small tree, and the opposite side against an adjacent tree at the beginning point of the bend. Very slowly and carefully, the opposite end of the ski would be drawn toward a third tree with a small rope until the bend was slightly greater than that desired, thereby allowing for some spring-back when the tension was released. See Fig. 17

Once the wood had thoroughly dried, the rope was removed and the ski was ready for finishing. Straps from an old leather belt were attached to the side of the ski at the front of the footrest, and the ends of a screen door spring, shortened to the proper length, were attached at the same point as the straps. A notch was cut in the heel of each boot into which this spring,, under tension, could be placed. With the toe of the boot under the strap, somewhat of a quick- release device was the end result. See Fig. 18

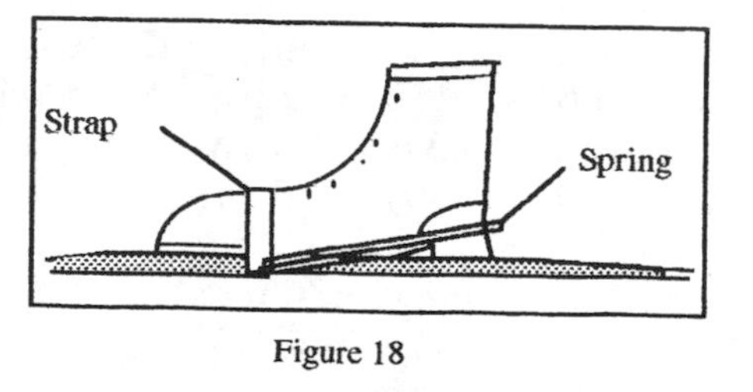

Figure 18

Detail of Binding

The bottoms of the skis were sanded as smooth as possible, and coated with paraffin wax. The remainder of the surface was coated with lard to help prevent water penetration.

Downhill skiing was excellent on the sleigh run by our house, on the hillside between the Matsons' house and the golf course, and also on its first, second, third, and ninth fairways. Cross country skiing was good on the other golf course fairways, and on the ridge area just to the west of the golf course in the vicinity of the baseball diamond, both to the north and to the south. Some of us became very proficient skiers on our homemade skis; Harvey Hurley so much so that he was assigned to the Army ski patrol when he entered the service during WWII. During my last winter in Tererro, when Joe was away to school and I was without a job, I would spend most of my spare time skiing on the golf course.

Although one could occasionally skate on the river in selected areas where the water gentled and froze without a rippled surface, the lake at Irvin's Ranch and the south lake at Nunn's cabins (now known as Tres Lagunas) were favored spots for skating. Most was done at night when several old tires were set afire to provide both warmth and light Almost every Friday or Saturday night from mid-November to late March, two or three who could drive (in those days driver's licenses weren't required) would borrow their parents' cars, drive around town gathering up a load of friends, and off we would go for a skating party and wiener/marshmallow roast at one of the lakes or another. The Hurley children's grandparents were caretakers at Nunns; we could always get permission to skate there; Irvin was not always quite so accommodating.

For the most part, we all owned clamp-on skates in our younger years, but graduated to shoe skates when our parents decided our feet would not outgrow a pair in just one year. I was skating at Big Bear Lake in California on December 7, 1941, using the pair I received for Christmas in 1936, when the news that Pearl Harbor had been bombed was announced over a loud speaker.

At the mill, a pond which had formed below the tailings dam froze in the winter. This is where those living there skated, but they also would meet a group of us at Irvins or Nunns on occasion. They also had the benefit of a place to roller skate in the summer. The railroad boxcars, used to bring coal to the mill for operation of the power-generation plant and to transport the partially refined ore out to smelters, often sat empty for several days. Since they were parked on a siding away from the mill loading dock, and away from the prying eyes of employees, the kids would use the cars as roller skating rinks. A kerosene lamp hung at one end was their light source. They thought their parents wouldn't know where they had been, but the tell-tale coal dust on their clothes always gave them away.

The one diversion missing in Tererro was a place to worship—there were no churches. The closest Catholic church was at Macho Canyon, but one had to go all the way to Santa Fe or Las Vegas to attend protestant churches. As a result, the great majority of the Anglo children grew up without a religious background, as did many of the Hispanics. I can recall only one summer when some of the ladies conducted a Bible school for the children. It was poorly attended and wasn't tried again.

During my growing-up years, I can recall having been taken to church only two times. When Joe attended New Mexico Military Institute, he was required to attend chapel services each week, otherwise his religious background was little different from any of the rest of us living at the mine. For those living at the mill, the Santa Fe churches were much more convenient if folks wanted to drive the approximate 45 minutes to attend church; for those in Tererro, the drive was almost two hours.

This lack of religious opportunity, in a sense, made hypocrites of all of us Boy Scouts. We pledged ourselves to be reverent and to "...do my duty to God and my country", but we knew little about God and what it meant to be reverent. In any event, it made little difference for Joe and me, for we both became very active church members after we were married. LaVoyse married a Baptist minister, but Delta never became a religious person. How it affected others in later life is not known.

chapter 24

* *

Tragic/Near Tragic Events

Injury and death, through accidents or violence in its many forms, and near tragic events were no stranger in the community. There were many brawls resulting from either drunkenness or racial disagreements, or both, and these often led to knife stabbings or injuries from broken bottles used as weapons. Many went unnoticed by the general population, even the one which led to the death of a mucker named Leslie. He was referred to as a red-neck and apparently disliked Hispanics intensely since he was constantly goading them into fights with himself. Quite husky and weighing over 200 pounds, he usually came out the winner, but not when he picked on Jesus Mildenado one evening on their way home from work.

Jesus, about five feet in height and weighing perhaps 130 pounds, had nine children, all under twelve years of age; he was not only a good father, but a good citizen as well. According to reports, he didn't drink and, although he was a Mexican national, he always wanted to carry the American flag at the head of the Mexican Independence Day parade held each year in Pecos. He wore a black sombrero which he always tipped to the ladies.

To protect himself when Leslie, about twice his size, picked the fight with him, he pulled out his knife and stabbed him in the heart. He picked up his black sombrero, sat it squarely on his head, and went home. Someone carted Leslie off to the hospital, where he died later that night. Jesus went back to work the next day as if nothing untoward had ever happened. Leslie was buried in the local cemetery just to the south of the salt lick. His was the first of six burials there.

County Sheriff Miller came that day to investigate the death, fully expecting to arrest Jesus before the day was over. His investigation showed Leslie to have been a trouble maker and Jesus to be a peaceful, law-abiding citizen. He talked to Matson about the incident and, in essence, said that probably Leslie would have been killed by somebody else sooner or later. He reasoned that if Jesus were arrested and sentenced to prison, his wife and children would have no means of support and, as he put it, "....the Ladies Aid Society will be after the County to support them." He suggested that the incident be considered a justifiable homicide and the matter dropped, thereby saving the County the cost of a trial. Matson agreed and Jesus continued working at the mine until it closed.

Three years earlier, Henry Pick was murdered on the Pecos highway near Dalton Canyon. Few residents of Tererro ever knew about the incident in which Jesus killed Leslie, but everybody knew about Pick's murder, and even today, some 60 years later, it is still a topic of much discussion and speculation among those who once lived there. The murder was never solved, and there have been numerous speculative accounts as to what might have happened.

In an earlier chapter, it was mentioned that Henry always came to the mine from Santa Fe on the day before payday to bring a supply of cash. One day, he was found slumped over the steering wheel of his car with a bullet in his head. His car was up against the bank on the wrong side of the road and had scraped the bank for several yards before coming to a stop.

The following account of the incident was related to me by Herbert Pritz, a truck driver for the company, who was certainly the first person to come upon the murder scene:

> I had been to the mill to pick up a load of supplies for the mine and had started back about 10:30 am. At the point where the mill road entered Highway 85, a black Chevrolet was parked at the side of the mill road facing toward the town of Pecos. Two men in black overcoats were standing on the driver's side of the car facing toward the mill, and a third man lay on the ground as if he were either putting chains on, or taking them off the left rear tire. However, there was no mud or snow which would have made chains necessary, and my curiosity was aroused as to why they were there. Since the two standing men faced toward me as I drove by, I got a good look at their faces.
>
> About thirty minutes later as I started down the La Posada hill (about six miles north of Pecos), two cars passed me at a high rate of speed. Two men were in the front seat of the trailing car, and another sat in the rear. I immediately recognized it as the car I had seen on the mill road, and the two men were in the front seat as those who had been standing beside the car. The trailing car was not more than thirty feet behind the lead car, and seemed to be trying to pass the one in front. Traveling at such a high rate of speed, both cars were soon out of sight with the black Chevrolet still trailing the lead car.
>
> Just beyond the first Pecos River bridge (seven miles north of Pecos), the road makes a rather abrupt bend to the left toward the mouth of Dalton Creek canyon. At this point, I met the Chevrolet and its three passengers coming back traveling at the same high rate of speed. As they came around the bend, the speed carried them over into my lane of traffic, and I had to veer to the right edge of the road to avoid being side-swiped. The interval between the time the two cars passed me going north, and when I met the car coming back south, was about six minutes.
>
> As I drove along and attempted to make some sense as to what this was all about, no more than 400 yards further along the road, I came upon the lead car of the two lying against the bank of the mountainside on the left side of the road—the river flowed on the right side. I saw the driver slumped over the steering wheel of his car looking quite dead. I thought of turning around to go back to investigate, but to tell the truth, I was scared and decided to drive

to the mine as fast as I could to report what I had seen. I told Billy Montgomery, and he phoned the County Sheriff's office. He was informed they already knew about the murder and that sheriff's deputies and the State Police were at the scene of the crime. They said it was Henry Pick slumped over the steering wheel, and that he had been shot in the head at close range.

Apparently the next person after Pritz to come upon the scene of the murder was a salesman from Gross Kelly Mercantile Co. who recognized Pick and drove back to Pecos to report what he had discovered. He found a shoe box on the seat beside Pick containing $5000 in various denominations of currency and coins. He took the box back to Pecos with him and turned it over to a sheriff's deputy. Later investigation showed that a bag containing tire chains was missing from Pick's car, and it was presumed the robbers had snatched them from the back passenger compartment of the car thinking that it was a bag of money, and had thought the shoe box really did contain a pair of shoes. That, of course, is pure speculation as to what actually occurred.

In any event, there was very little time for the three to do much more than grab what was in sight, and to get away from the scene since they knew full well that Pritz would be driving by very shortly, and they had no idea how soon someone might approach from the opposite direction. Pritz believes that they could not have left the scene of the crime much more than one to two minutes before he met them headed back down the Pecos highway. He was interrogated by the State Police on several occasions, and was taken by them to Las Vegas to view possible suspects in line-ups, but was never able to identify anybody as one of the three he had seen on the mill road.

Within a few weeks after Henry's death, Emil married his widow and moved in with her in her Santa Fe home. The story that went around was that it was an old Jewish custom for a single brother to marry the widow of a deceased brother so as to care for her and any of the brother's minor children—a small son in the case of Henry. This story was generally accepted as the reason for the somewhat hasty marriage.

There was a great deal of speculation—there still is—as to who might have committed the murder. One rumor had it that Emil had hired a killer to do the job. A good many years later, Emil committed suicide, a fact that revived the rumor, but nobody really knows for sure what happened, and never will.

My parents were close friends of Emil and his younger brother, Marcel, both of whom were guests in our home many times before the tragedy occurred and before Emil married and moved to Santa Fe. Marcel eventually married, and his wife also became a good friend of my parents. I worked in Picks' store during the 1938 Christmas season and, when the mine closed, Marcel wanted me to help him drive with his wife and small son to the New York World's Fair, in return for which he would pay my expenses. I had already made firm plans to move to California to attend school and had to refuse.

After the mine closed, Emil and Marcel purchased the White Swan Laundry in Santa Fe, and when I would return there on vacation in later years, I never failed to stop by the laundry to visit with them. I personally don't believe Emil had anything to do with the murder. Henry's son at the time of the writing of this history is the mayor of Santa Fe.

There is another interesting story about the two brothers which Emil told my parents. Emil had a large scar on his left cheek from a wound he received while serving in the German Army during WW I. Henry had already migrated to the United States and was serving in the American Army. On Christmas of 1917, there was a short truce in the fighting. Troops from both sides came out of the trenches, and walked over to wish their enemies a Merry Christmas. Although the brothers were Jews, they participated in the greetings, and by sheer coincidence found that they were in the trenches opposing each other.

There was a deputy sheriff, easily identified by the patch he wore over one eye, assigned to the Pecos/Tererro area by the name of Felix Vigil. Some white folks in the area thought that he was particularly intransigent, and that he hassled them far more than he did the Hispanics when they became drunk and disorderly. He was killed one night when his car failed to make the Gables Lodge curve just north of the fish hatchery. It was rumored, but never proven, that some Anglo men had forced him off the road.

When I was about twelve, one evening about dusk Mom sent me down to the garden to fetch a head of cabbage. The cabbage grew on the lowest side of the garden next to the fence. I cut the selected head, and rather than walk back to the gate to exit the garden, I crawled through the barbed wire at that lower end onto a trail leading to the five bunkhouses situated on the hill above the garden. Just as I stepped on the trail, a young Anglo man ran down the trail past me as fast as he could run, then turned onto the trail leading to the mine. Just as he turned, probably no more than fifteen feet from where I stood, three shots rang out from the area where the bunkhouses were located.

I still vividly recall hearing the bullets slam into the earth near me. One ricocheted off a rock and went screaming into the woods behind. I hadn't the slightest notion what was occurring, and only became frightened when a young Hispanic with a pistol in his hand came running by and asked which way the Anglo fellow had gone. I rushed back to the house, leaving the cabbage behind, to tell my parents what had happened. Dad immediately drove down to tell the local peace officer (we called him the sheriff, but officially he wasn't). His investigation revealed that the Anglo man had been having an affair with the Hispanic's wife, he had found about it, had beaten his wife, then came looking for the interloper at the bunkhouses, his loaded pistol at the ready.

The intended victim happened to see him coming in time to escape down the trail. Had he been a minute or two later, he probably would not have escaped the husband's wrath, and would have been killed on the spot. The husband was incarcerated in the local jail overnight. In the meantime, the intended victim hastily left town and probably never returned. The husband was let out of jail the next day since there was no one to press charges against him for attempted murder.

A party in celebration of the abolition of prohibition in the spring of 1934 resulted in the death of Mrs. Hoag, wife of the mine superintendent, and the serious injury of a young woman, Jody Kleinsworth. Prohibition had been abolished in December 1933. The party was held at Viles Ranch in Cowles and was attended by many office employees, their wives or girl friends, a couple of teachers, and sundry other individuals from the town, all of whom were not averse to more than an occasional alcoholic drink. Around midnight the party was getting pretty raucous and several had had a few too many celebrating drinks.

Mrs. Hoag wanted to go home, but her husband, Cliff, was not yet ready to depart. Jack Finlay, the young man Matson had hired early on at the request of Steele, volunteered to take her home with him and his date, Jody Kleinsworth. He also, unfortunately, had had too many drinks, but in those early days after repeal of prohibition, not much thought was given to the dangers of drinking and driving. The road to Cowles was narrow and winding with no guard rails on the sharp outside curves. About two miles south of Cowles, he failed to make the sharpest turn, and drove over the edge. At the point where the car left the road, the Pecos River flowed about 200 feet below, and the incline of the mountainside was at about sixty degrees from the horizontal.

The car was totally demolished as it tumbled down the incline and came to rest against a pine tree just above the river. Mrs. Hoag was killed instantly, Jody received a compound fracture of her tibia and was hospitalized for nine months. Finlay's injuries were somewhat minor, which some attributed to his inebriated state. Not long after the accident, he resigned and moved away. My father served as a pall-bearer at Mrs. Hoag's funeral in Santa Fe which all of our family attended. It was the first for any of us children.

When I was in the third grade, one spring afternoon there was a very heavy thunderstorm with many lightning strikes near the school house.[1] The students all became frightened and many hid under their desks. Our teacher, Miss Isabel Barrett, calmed us by giving us a quick lesson on thunderstorms, on lightning arrestors located on the roof of the school, and by explaining why we should never take refuge near a single tree when caught out during a thunderstorm.

That summer, with some scrap canvas, Joe and I built a play tent in a grove of five relatively large and several small pines below the house. As was usual, a thunderstorm came up, and we were caught in the tent when a heavy downpour began. Lightning struck one of the tall pines nearby, splintering the side of the tree facing the tent into dozens of pieces, one of which came ripping through one side and out the other, barely missing me. The heavy downpour did not deter us from beating a hasty retreat to the safety of the house.

As a result of that incident, a decision was made by the company that all the tall pines near any of the houses should be cut down and, before the summer was over, they were all gone except two. They were left standing because there was no direction in which they could have been felled in which they would not have damaged some structure or a power line. One of the two, about 100 feet tall, was relatively close to our house, and the other was immediately behind Joe's tent.

Miss Barrett's earlier lesson on lightning had not fallen on deaf ears—I remembered all too vividly how she had told us that it would strike the tallest object around, and for as long as we lived there, I lived in fear of that single tall tree being struck. I was certain that some large piece of it was going to fall on me through the roof

---

[1] Thunderstorms were very prevalent during the late spring and summer months. They came in the early afternoon in the spring and slowly shifted toward late evening by the end of summer, and one could be expected almost daily.

of our tent-house bedroom. I lost a lot of sleep during nighttime thunderstorms worrying about it, but it was worry in vain. The tree still stands sixty years later, and the small trees surrounding it then have now grown to be almost as tall. The other tree also still stands; on it is a memorial plaque for Joe, who died in 1987. Under it, and at other places where he and I played as boys, are his ashes, scattered there by his family and me.

One winter evening, Delta and I were busy at our daily chore of splitting wood and kindling for the wood-burning stove. Joe had come over to help me carry the finished product into the house. Jack McCloud, son of the man who did the diamond drilling for the mine, had come home from school with Delta, and was placing the wood to be split onto the cutting block which Delta would split with a double bladed axe. Jack, thinking that he had not properly placed one of the pieces of wood, reached out with his foot to move it into better cutting position. At that moment, Delta swung the axe with a mighty blow aimed at the original location of the piece of wood. Axe and toe came to the same spot simultaneously.

Because there was snow on the ground, and the temperature was below freezing, Jack was wearing a pair of light cotton socks, a pair of wool socks, fleece lined boots and a pair of heavy galoshes. Too late to forestall total tragedy, Delta saw what Jack had done and tried to slow the momentum of the axe head. Fortunately he was able to slow it enough that the blade only sank into Jack's foot as far as the bone in his big toe. Dad arrived home from work about that time, placed a tourniquet on Jack's leg and drove him to the hospital where Dr. Smith spent an hour or so sewing up the damage to the tops of three toes. Had Delta not tried to impede the speed of the moving axe, Jack would surely have been minus three toes, although with Dr. Smith's skill at emergency surgery, he might have attempted to sew them back into place. Success at that kind of surgery was to be many years in coming

My folks decided to plant a small clover lawn area near the kitchen entrance to our house. Because Littrell's dairy cows roamed the company property freely, they discovered the clover, and would leave the lawn in shambles. Delta and I were assigned the task of stretching a three-strand, barbed-wire fence around the area to keep them out. As usual, Joe was there to help. We worked all morning and managed to stretch the top strand, using some smaller pine trees around the area as support for the wire rather than setting fence posts.

It was Mr. Matson's custom to whistle for Joe when it was time for him to come home, and he did so that day when he arrived home for the noon meal. We were sitting on the steps leading up to the kitchen, admiring our morning work, when he whistled, and that whistle meant that Joe was to waste no time in coming home. He jumped up from where we sat and began his mad dash home. Unfortunately, he forgot that there was now a strand of barbed wire blocking the path he usually took, and he slammed into the wire going full tilt. He saw it just in time to turn his head to avoid injuring his face, but one of the barbs caught his right ear where the lobe joins the jaw and practically tore off the lobe. Another quick trip to the hospital and another delicate stitching job for Dr. Smith. Mrs. Matson was furious that my parents would have allowed it to happen. We removed the top strand of wire; the lawn and the fencing project were abandoned.

I learned early in life that an unloaded gun is not necessarily so. I owned the only bicycle in town and used it to deliver the *Santa Fe New Mexican* newspaper, although I pushed it up a lot of steep hills.

Al Colbert, son of one of the teachers, had a small, single shot 22 rifle. He was going to spend Friday night with me, and we were going rabbit hunting the next day. When I delivered his mother's paper that Friday afternoon, he came along on the remainder of my paper route, bringing along the 22 rifle. We were at the tail end of the route in Rico Town; he was riding the bike. I tossed papers from the bag hung over the rear fender and carried the 22.

We came to the home of Felix Lujan, a very likable Hispanic boy. We called him Felique or Colorado (the Spanish word for red) because of his red hair. He was perhaps two years older than the two of us, and somewhat bigger. He wanted to ride my bike; but, since it was growing late, I refused to give him permission. He decided to wrest the bike away from Al anyway, and I foolishly pointed the rifle toward him and told him to stop. I had no idea that it was loaded, and I pulled the trigger without thinking. It was just dusk, and because the rifle barrel was not clean, a blazing bullet skimmed directly over his head, missing by not more than an inch.

I think all three of us must have turned deathly pale as we contemplated how close I had come to killing him. We didn't go rabbit hunting the next day, and we never told our parents why. I never handled a gun again until one was put in my hands when I joined the army in 1944.

Strangely enough, on Christmas Eve of that year, I was on a one-day pass from Ft. Belvoir in Washington D. C. I was in the Pepsi Cola Canteen going up a stairway to the second floor when I bumped into Felique, also in the army, coming down the stairs. We had a Pepsi Cola together and talked about the incident and our Tererro days. We exchanged addresses and parted, promising to write on occasion. About two months later, I received a letter from the Red Cross telling me that he had been killed in the Battle of the Bulge. Included in the envelope was an unfinished letter he had started to write to me. They found the slip of paper with my name and address in his belongings.

In July and August, the river water warmed enough that we could occasionally swim in it, usually skinny-dipping, at the only place where the water was deep enough to dive and actually swim. The water cascaded off some huge boulders into a mildly still pool. The configuration of the rocks and boulders on the lower side of the pool caused the water to circle around into a whirlpool about 12 feet in diameter. We could swim straight across or in a circle, either with or against the direction of flow.

The pool was situated about halfway between the twin-bridges meadow and the dairy, so our trips to it were seldom, and always made with considerable forethought based on how warm the weather was. Occasionally, Joe and I would go with other boys, but usually it was just the two of us. The summer we were 12, we convinced Pat Montoya, nephew of Matson's maid, to go with us. He was perhaps two years younger and weighed about 20 pounds less than we did. When we got to the pool, he was very reluctant to go in, but we finally convinced him that a good dive would put him almost to the opposite side, where one of us would be positioned to help him out.

He dove in, but came up right in the vortex of the whirlpool, and he could not swim well enough to extricate himself from the swirling water. I went in to help him, but was not prepared for his panic-stricken survival instinct. Normally, somewhat of a weakling, he became a tower of strength and would not cooperate in allowing me to put my arm around him and swim to safety. He simply continued to force me under as he strove to keep his head above water by attempting to climb on my shoulders.. As a result, I was the one who almost drowned. Joe found a pole which I finally grabbed, and he pulled the two of us out of the vortex into shallower water. I had ingested far more water than Pat had, both in my stomach and my lungs, and was a long time in recovering enough that we could dress and go home. After that experience, I became fearful of being in water with other people in close proximity to me and probably Pat never overcame his fear of water.

Herman Morse, Norman Littrell, and Delta all went to work as mine timber rustlers at the same time. Herman smoked a pipe. While scuffling with somebody one day during the lunch break, the pipe was accidentally bumped, and the pipe stem driven into the back of his mouth. He didn't go to the doctor for treatment and the wound became infected. Dr. Smith was unable to cure the infection, and he died in just two days. It was a shock to all of us who had grown up with him, for he was only 18 and the first of our group to die.

My sister, Balpha, was the next. Her marriage at age 16 had been a mistake, for she turned out to be a victim of what is known today as the "battered wife syndrome." She left Edwin on several occasions because of the beatings, but always returned. She divorced him in 1934, discovered she was pregnant, and a few months later, they were re-married shortly before the birth of a son, DeWayne.

Soon after, Edwin took a job as a maintenance mechanic at the Montezuma boarding school she was attending when they met, and they moved to the campus. I visited them during the last week of December, 1938; while I was there, she received another beating. She pleaded with me not to interfere, so I went back home to Tererro. The next week she left him for the last time.

She moved in with a friend in Las Vegas who had a ten-year-old daughter. The three of them went to a movie on the evening of January 10 and were walking home on the sidewalk, three-abreast, with the daughter on the side away from the street. A drunken motorist came up behind them, failed to straighten his car out of a curve in the road, jumped the curb, and struck the two women. Both were killed instantly; the daughter was not scratched.

He stopped, got out of his car, looked at what he had done, got back in his car and drove away, but not before the daughter got a good look at him. The policeman investigating the accident was questioning the daughter, when she looked up and saw the man standing in the gathered crowd. She pointed him out and he was arrested for vehicular manslaughter. He had driven to his home only a block down the street, put his car in the garage, and returned to the scene.

He was the County Clerk for San Miguel County, and as I was researching material for this book, I came upon his signature on Amco's incorporation papers. He was sentenced to three years in prison, and was paroled one year later.

chapter 25

✲ ✲

Some Uncommon Folk, and the End of An Era

Other than Mr. Matson, perhaps the most noteworthy and highly respected resident of Tererro was Dr. Warren George Smith. He was fluent in Spanish and was regarded in high esteem by the predominantly Hispanic population, as well as the Anglos.

Dr. Smith was faced with many unusual emergency medical situations due to injuries sustained in mining accidents. He became an extremely proficient orthopedic surgeon without benefit of special education in that branch of medicine. His use of the silver quarter as a silver plate in my head is testimony to his ingenuity in making do with whatever was at hand. However, when he became the company doctor, he was no stranger to performing surgical procedures under unusual circumstances.

He graduated from Cornell University in 1906 with a Doctorate of Medicine and a scholarship for further study in Florence, Italy. When he was ready to return home, he was short of funds and had to board a tramp steamer which took him to Vera Cruz, Mexico. There he was robbed of all his possessions, and he was unable to continue his homeward bound journey except by attempting to walk back to the closest town in the United States. And this he set out to do.

The Mexican revolution was in full-tilt at the time. En route to the border, he was captured by Pancho Villa's troops and forced to provide medical attention to women, children, and the wounded of the bandit's rag-tag forces. Some two years later, in Chihuahua, a couple of Villa's leaders with whom he had become friends, fearing that he was going to be put to death by Villa, provided him with a horse and provisions, so that he could escape to Texas. From there, he ended up in-and-around Las Vegas, New Mexico, and in 1911 worked as a sawmill operator nearby.

With the help of a homesteading rancher, Harry G. Arnold, who he treated for cancer, he obtained a license to practice medicine in New Mexico in 1912, the year that New Mexico became a state, and set up a medical practice in Las Vegas. Later, he was in general practice in Mora when he signed the contract with Amco as company doctor.

He married one of Arnold's daughters, Alma, in 1915. She became his assistant in the hospital, and the two of them became a first-class surgical team. In addition, she kept the hospital records and was busy as a mother with five children to care for. She was equally well liked by all.

In addition to his duties with the company, Dr. Smith was interested in politics, having served for two years in the New Mexico State Senate. He also owned several flocks of sheep which were pastured in the mountains to the east of the Pecos River.

One would not think of Jessie Gonzales, Matson's maid and cook, as a woman of much esteem. She was very quiet and unassuming, and had no close friends or social life. However that opinion would change after she and "Chief" moved to Santa Fe when the mine closed. Nobody knew why her husband was called "Chief," but the nickname proved to be prophetic. He got a job with the New Mexico State Police, and began working his way up in the ranks; eventually he became its chief, a job he held many years. Meanwhile, Jessie took it upon herself to get some advanced education, after which time she ran for and was elected as New Mexico's Secretary of State, a position she occupied for two terms.

Harold Walter came to work at the mine in 1929 as an accountant, and in his spare time became an avid amateur nature photographer and mountain climber. In the high country of the wilderness, there was no dearth of beautiful photo-opportunities, and he spent most of his spare time taking advantage of them. In 1930, May Lynch was hired as the fifth and sixth grade teacher in the school system. The kids, especially the boys, (I was one of them), fell in love with her, and so did Harold Walter. They were married during the summer of 1931. Harold had made arrangements with Matson for he and his bride to occupy the house which had been built for Mrs. Clark and her son Tom, since they moved to the mill that year.

Harold now had a beautiful woman to include in his photos and he gradually became an expert and highly respected nature photographer. Two years later, he was transferred to the mill, and May was hired as a teacher in their school.

When the mine closed, they moved to Santa Fe where Harold went to work for the State Highway Department and May became Secretary to Elliott Barker, State Game Warden. For several years, until he died from an untimely heart attack, Harold provided pictures for the cover of the *New Mexico* magazine and the rotogravure section of the *Denver Post*.

One day he was taking pictures atop one of the mountain peaks south of the Truchas Peak that was thought to be the highest in New Mexico at 13,102 feet. The day was crystal clear, and Wheeler Peak close to the Colorado border could be seen. As he looked through his camera which he had leveled perfectly, he could see both Truchas and Wheeler in his viewing lens and realized that Wheeler appeared to be somewhat taller than Truchas. He sent his pictures to the U. S. Coast and Geodetic Survey and suggested that the height of Wheeler might be more than shown on maps at the time. They re-surveyed and discovered he was right—Wheeler was 13,161 feet, 59 feet higher than Truchas, and future maps were changed accordingly. In 1958, an unnamed peak just south of Wheeler was name Mt. Walter in his honor.

May died in her sleep on October 28, 1992, at 87 years of age, still full of vim and vigor, and always in touch with the many students who had loved her for over 60 years. She loved to write poetry, especially amusing little pieces like the one she wrote on her 87th birthday. In 1988, she had a small booklet of her poems published that was entitled, *Just For Fun.*

ON REACHING EIGHTY-SEVEN
By May Walter

Four score and seven years ago,
On the 31st of May,
Ancient records will reveal
I first beheld the light of day.

My parents were delighted
'Til the country doctor said,
"Oops my friends, I'm sorry
But I dropped her on her head."

That could account for many things
That befell me through the years:
Vicissitudes and ups and downs
That brought both fun and tears.

But, somehow, I survived it all,
And folks have been so kind,
Not once have they reminded me
I have a feeble mind.

The announcement on May 20, 1939, that the mine would cease operation in just 11 more days came as a blow to the majority of residents, both at the mine and the mill; few were prepared for it. The Great Depression was just showing signs of recovery, but there were still few jobs to be found anywhere. For most employees, finding another job at another mine was foremost in their minds. They knew that those who arrived first would get any available openings, so there was a mass exodus of both towns beginning on June 1.

I happened to be in California visiting my brother at the time. I arrived home on June 6 to find my parents were not there, and with no word as to where they had gone. In those days, nobody locked their doors, so I simply walked in. There were only a few clothes missing from the house, so I presumed they had simply taken a short vacation somewhere. Later I learned that since I was not expected home for another week, they felt no need to leave information as to where they would be. There was little to eat in the pantry, so I headed downtown to purchase something.

Much to my surprise, there were no stores open and very few people were about. I went over to the office in search of Mr. Matson to find out what was happening, and to learn if Joe would be returning home for his birthday the next day. He assured me that Joe would be home, and then explained about the mine's closure. In addition, he told me about a violent hail storm in Las Vegas a few days earlier in which many thousands of window panes had been broken by the golf-ball-size hailstones. Dad had answered a call for men who could glaze windows, and would return when the emergency was over.

Mom, Dad, and La Voyse returned about a week later, packed their belongings for pick-up by a mover, and headed for Carlsbad. Dad, was the mine's only full-time carpenter at that time, and knew there would be no competition from Tererro men for any carpenter opening at the Carlsbad potash mines. He was right, for within a week he had found a new job.

A couple of days later, Mr. Matson asked Joe and me to come to his office as he wanted to talk to us about a summer job. He proposed that the company turn over to us the company garage and its two gas pumps, so that we could operate a service station for the benefit of those few people still remaining in town and for the few company vehicles still in use. There was also the potential business from summer residents in the area and from itinerant motorists.

The company would start us off with both storage tanks filled with Conoco gasoline—white and ethyl—and a few cases of motor oil. At the end of the summer, we could pay for the gas and oil out of our profits, but we would be entirely on our own to make it a profitable venture. He insisted that we had to keep a good set of books to insure that we would not get ourselves into a losing situation. At the end of the conversation, when we had agreed to his offer, he said to me, "If you boys do a good job of this, I will have a proposition for you at the end of the summer." He wouldn't explain further but left the carrot dangling.

The following day we were in business, but we had more people asking us to repair flat tires than were buying gas. It seems that roofing tacks were falling on the highway from the trucks moving the scrap lumber and roofing paper to disposal sites, and they were a real hazard to anybody driving on the highway. We painted a sign on an old tire "Flats Fixed" and placed it in front of our station. We set a price of $1.00 per tire, and business boomed. Mr. Matson came by and asked when we were going to put up a sign saying that gas was for sale. We never did—the pumps themselves were our advertisement. When the books were totaled up at the end of summer, we had made more profit fixing flats than we had on the rest of the business, and I had saved about $100.

Delta had worked as a timber rustler at the mine for a year after graduating from high school, and in 1938 had saved enough for his tuition to attend Aero Industries Technical Institute, an aircraft trade school in Burbank, California. He enrolled for the six-month course and, upon graduation, secured a job at Lockheed Aircraft Corp. from where he retired 35 years later as superintendent of their ship building operation in Seattle. He married a Kansas girl in 1943, and they have two children.[1]

In early August, 1939, Mr. Matson again asked me to come see him. He told me that he was very pleased with the way we had operated the service station, as I had shown I could assume responsibility. He said that, if I would like to attend the same school Delta had in California, he would loan me the money at no interest, and I could repay him after I got a job. It was for me, at that time, the most wonderful thing that had ever happened, and I accepted his offer.

He paid all my expenses, including transportation to California, until April 7, 1940, when I graduated and went to work for Vultee Aircraft Corporation in Downey, California. He did not have me sign a note, but loaned me the money on my word that I would pay it back. The total debt amounted to over $1,200. My hourly income that first year was 55¢ per hour, but I did work a lot of overtime. I made $1,298 the first year of

---

[1] For their 50th wedding anniversary gift, I gave them a hand-bound copy of the then completed portion of this book, for he had a rare form of cancer and the doctors could not predict how long he would live. He died on May 12, 1993, two weeks and three days after their anniversary. Dad had died in 1972, and Mom had died in 1987.

employment, and almost $1,800 the second year. I was able to pay off the last of the debt about three years after I went to work.

La Voyse finished high school in Carlsbad in 1940, then attended college in Las Cruces where she met and married Howard Russell, another student, just four months later. He eventually became a Baptist minister, and they have three children.

A large segment of Tererro's population went to Carlsbad, where other mining jobs were available, and most found them sooner or later. Others moved to Las Vegas, Santa Fe, and Albuquerque to find different kinds of employment. Some moved to Montana, Wyoming, Idaho, Oregon, Arizona, and California.

Delta had written a number of his friends about Aero Industries Tech and had told them how graduation almost guaranteed a job, since the school was sponsored by the six aircraft manufacturing firms in Southern California. One friend and fellow timber rustler at Tererro, Al Kangas, enrolled as soon as he heard the mine was closing, and also went to work for Lockheed upon graduation.

About nine months before the mine closed, Mr. Hurley, a tramway maintenance worker, was doing some welding and his torch exploded a blasting cap someone had dropped in the dirt where he was working. His leg was seriously injured, and he was placed on partial disability. When the mine closed he received a final settlement for his injuries, and moved his family to Wichita Falls, Texas. He invested in a bowling alley hoping to have a business in which his two sons, Harvey and Jimmy, could work. Unfortunately, it turned out to be a poor investment and he pulled out. He had enough capital remaining to send both boys to Aero Industries Tech, so they moved to California too, and the boys enrolled. After graduation, they went to work for North American Aircraft Corporation.

About two weeks after arriving in California and before their boys started to school (all of us lived in Glendale), the Hurleys asked Al Kangas, Delta, and me over for dinner one evening. Mr. Hurley said he had something he wanted to tell us. He told us that before coming to Tererro, he had been convicted and served time for some felony in Texas. Upon moving to Tererro, they had changed their name from Hicks to Hurley, to insure that Amco would not find out about his past. All of the three children's birth certificates showed their name as Hicks, and to enroll in Aero Industries Tech, or to work in the aircraft industry, one had to show a birth certificate or naturalization papers as proof of American citizenship. Their secret was about to be revealed, and he wanted us to know in advance. To us it made no difference, but it was difficult to start calling them Hicks after knowing them as Hurley for the previous nine years.

Omer Tucker decided to join the Marines and was accepted. He was given a date to report to Camp Pendelton in California. He had a motorcycle on which he departed about two weeks before his induction, expecting to stop by Farmington to bid his high school girl friend good-bye. All of the Tucker children had gone there to live with Grandma Tucker and attend high school. He and his girl friend decided to get married before he left Farmington. When he arrived and reported that he had a wife, his inductance was canceled and he returned to Farmington. He became involved in real estate and eventually was elected mayor.

Our scoutmaster Dan Rogers and his wife, Mable (a former school teacher), moved to Dallas where Dan went to law school, and eventually opened his own law practice.

Several of the young men joined the Civilian Conservation Corps while others went into military service.

In the very early 1900's, an Hispanic couple by the name of Antonio and Juanita Salazar lived in a cabin in the Tererro area, and Antonio worked on one of the ranches, either the Simmons, the Chapin, or the Cornell ranches—it is not known which. On February 9, 1909, in that cabin, Juanita gave birth to a daughter, Francisca. Eulogio Padilla, a handsome, blond Spaniard, came to the mine to work sometime between 1926 and 1936, where he met and fell in love with Francisca, 23 years his junior. They were married on August 26, 1936 in the Catholic church in Pecos. Francisca's brother, Miguel, also worked at the mine as did Jack Mares, brother to Miguel's wife, Elena.

On May 12, 1993, I happened to look in the obituary column of our local paper and read that a Francisca Padilla, born in Tererro, New Mexico had died in a nearby town, and that she had a surviving daughter, Rose, who lived in my own town. I contacted Rose, who had been born in Pecos in 1940, and she provided me with documents attesting to the above facts.

This one-in-a-million chance of encountering a total stranger whose mother had been born and lived in Tererro, reminded me of another similar occurrence in 1956. I visted a small manufacturing plant (a customer of the company for whom I worked) and was having a conversation with the owner. We were discussing where we both came from—he fom New Jersey—and when I said New Mexico, he asked where? When I said that it was a town he would have never heard of, and then said Tererro, his response was that he knew it before it was a town. The Simmons ranch owner was his uncle, and he had spent several summers while in college working on the ranch.

Of the five who started in the first grade with me in 1926, Joe Matson served in WW II in an armored division, then graduated from the University of Colorado, and eventually became Senior Process Engineer of General Motors' aluminum diecast plant in Bedford, Indiana. He married Lilah Lohnes from Pekin, Illinois; they had five daughters and a son. He died on March 30, 1987.

Dick Smith joined the Army Air Corps, was shot down over Germany, and held a prisoner of war for two years. He stayed active in the Air Force Reserve and returned to active duty during the Vietnam war. He retired as a Lt. Colonel. He and his wife had five children and he died in 1988.

Carl Phillips was drafted into the army during WW II, and upon discharge, went to work for a company doing business at Los Alamos. His wife's name was Claudia, and I have not ascertained whether or not they had children. He died in 1986

Tom Clark joined the New Mexico National Guard in 1938, and his unit was the first to be federalized shortly thereafter. They were sent to the Philippines where he was captured by the Japanese and died in the Bataan death march. Three young engineers, who had worked at the mine, had gone to the Phillipines to work in some mines there. They too were in the death march and did not survive.

I have not learned what became of Jane MacDonald.

As for me, I worked for Vultee Aircraft Corp until 1944 when I went into the service, serving nine months in the Army Air Corps, and 21 months in the Combat

Engineers. While stationed at Ft. Belvoir, Virginia, I met a Kansas girl working in Washington D. C., Lois Warren, and we were married two weeks after my discharge from the service. I attended the University of Southern California from 1947 through 1951, and we then began our family, having four children in less than three years, the middle two being identical twin boys. The oldest, Russell, and his wife, Tobae, have two sons, Daniel and Mark. Kevin and his first wife, Shelley, have a daughter Erin. He and his second wife, Lori, have a son, Ender. Kenneth and his wife Karen have a daughter, Karisa, and a son, Ryan, and they have adopted an orphan from Romania, Jonny. Robin, our daughter, has no children. I retired in 1983 after working 17 years as Deputy Base Civil Engineer at Travis Air Force Base, California. During the past 12 years, I have been actively involved as a volunteer in the work of Habitat for Humanity.

Only a few post-Tererro lives have been brought into focus to illustrate that, in general, we were not deprived by our having lived there, rather that our lives may have been enriched by the experience. Most of the adults and many of those who were children there have died, but those of us who can, gather in a yearly reunion in Pecos to celebrate the good times we had growing up together.

The mine operated and continued to grow throughout the depression. Although the wages paid were not outstanding, those who worked at the mine and the mill were able to support their families in a manner far superior to millions of Americans during the depression years.

Nobody lived in luxury, few ate the best of foods except on rare occasions, but none ever went hungry for want of something to eat, as so many others did then and are still doing today. As with many of the families, we had pinto beans as a pretty steady diet, but we, as did others, raised chickens and there was usually roast or fried chicken for a Sunday meal. Only on one occasion did we ever have turkey, and that was when Balpha's boyfriend took the family for Thanksgiving dinner at the messhall where he worked.

Three of my parents' brothers and two nephews came to seek employment, and all were hired. Their incomes far exceeded that which they had been eking out on their "dust bowl" farms of Oklahoma. All resided with us until they could find or build their own living quarters, but all eventually moved on when they found the cold winters intolerable.

It was only fifty years later that I began to realize the significance of the real world drama that unfolded around me during the thirteen years, three months, and seventeen days that Tererro was my home. Nor did I consider the fact that only Jimmie Russell and my father had lived there longer than I had. During those years, the population had grown from 192 male employees, two women and four children to over 850 employees and a bustling community of almost 3,000; in just three months, it had withered away to a half dozen, only one of whom, Jimmie Russell, had witnessed the whole drama, while I had witnessed all but the first four and one-half months. My father had witnessed all but the last three months..

Frances Earickson Slade wrote these words to me, "I think it would be marvelous for you to write a book about that rowdy, dowdy, dirty little town that really was so much fun!" Thisisthatbook.

The End

EPILOGUE

✲ ✲

In the 1920's and 1930's, there was no scientific or medical data identifying lead as a potential carcinogenic. It was used widely in batteries (and still is), in lead-based paint, as a solder, as lead pipe, as a material from which many small toys were molded, and many, many other applications too numerous to name. There was simply no concern about the applications to which it could be put to use.

After the passage of the first modern federal environmental statues in the early 1970's, AMAX (successor to the American Metal Company, Limited) was one of the first companies to form a corporate environmental group to coordinate necessary responses to the laws and to educate operating units of the company regarding compliance with the new requirements. Additional environmental laws have been added so that now virtually every aspect of a mining operation is regulated, and detailed environmental planning and government approvals are required before a mine can be developed. Such was not the case, of course, when the mine at Tererro was developed and the mill and tailing ponds started up near Pecos. Similarly, no requirements were in place when the operation closed on May 31, 1939.

As noted earlier, the mine has been secured by sealing the shaft and other openings and, in addition, the spillways of the two tailing ponds were left intact as it was believed that they would pass stormwater without damage to the tailing dams. In fact, when engineers from AMAX's environmental group inspected the site in 1986, after receiving a notice from the United States Environmental Protection Agency (EPA), the spillway on the upper tailing pond, which holds 85 percent of the tailing, was found to be functioning exactly as it should—by-passing stormwater around the dam. The lower tailing pond was a different story. The spillway had apparently failed and there was a large erosion channel through the deposited tailing which had cut completely down to the original surface of the streambed. A third small earthen dam had been constructed just downstream, apparently to contain the tailing material being eroded from the tailing pond. This third containment had completely filled and it, too, was beginning to erode.

The mill and power plant were, of course, completely gone except for concrete foundations and some scrap, and all buildings that had housed workers and their families as well as offices and schools had been completely removed.

At the mine site there remained a waste rock pile that held the barren and below ore grade rock that had been mined in the process of developing the ore deposits. The mine site appeared to be causing little environmental problems except for a couple of small areas of vegetation, which were showing some adverse effects. A healthy looking wetland area had developed in the area between the Pecos River and Willow Creek where the school had been located, and this area was directly below the rock storage pile. Beavers had dammed Willow Creek in several places, thereby creating the wetland area. Except for a few concrete foundations where the crushers, the tramline loading station, and other mine facilities had been located, there was no visible evidence of the town of Tererro or the buildings associated with the mining operation. Because the original notice from EPA addressed only the tailing ponds, and because the mine site appeared to have minimal problems, AMAX engineers turned their focus to the planning of remedial activities at the tailing ponds.

Some may wonder why a company which was only a successor to another company, which in turn owned only 51 percent of the operation during its lifespan, could be forced to come back and do work on a site which it no longer owned 60 years later. The answer lies in the federal Superfund law enacted in 1980 which makes liable for "release of hazardous substances" both current property owners and those who owned the property when the "substances" were "released." Since the State of New Mexico had acquired ownership of the properties, as previously described in Chapter 15, the State, too, is responsible under the Superfund law.

AMAX immediately engaged a large engineering firm with experience in tailing pond construction to evaluate the safety and stability of the two tailing ponds, and to investigate alternatives for handling the water flow to prevent further erosion. A period of protracted discussion and negotiation ensued involving both the New Mexico Environmental Department and the New Mexico Department of Game and Fish which managed the lands for the State. It was not until the end of 1992 that an agreement was reached with the State to allow a remedial plan to be carried out at the tailing ponds, and to begin studies and eventually remediate the mine site. This agreement provided that AMAX would pay 80 percent of the cost of remediation and the State would pay 20 percent. By then, AMAX and the New Mexico Environment Department had agreed that the best solution for the tailing ponds was to construct a lined channel some 70 feet wide and seven feet deep across the surface of both the upper and lower tailing ponds, to construct buttress fills against both dams, remove the tailing from the catch dam below the lower tailing pond, refill the erosion channel, and then cover and re-vegitate all the exposed tailing.

As the discussions and negotiations between the State and AMAX were proceeding regarding the tailing ponds, new concerns were arising concerning the mine waste pile at Tererro. Some fish kills had occurred at the State's Lisboa Springs Fish Hatchery located about two miles north of the Village of Pecos. Since the springs cannot provide all the water required by the hatchery, some water is also taken from the Pecos River. Although the exact cause of the fish kills still has not been determined, some believed that it was due to rain or snow melt flushing through the mine waste pile some twelve miles upstream, even though no fishkills had been observed in the river. Concerns were also raised about the possible health effects from mineralization in the waste rock. Even though miners still today mine lead ore deposits with no ill health effects, problems that have developed with children exposed to lead paint and leaded gasoline emissions have given rise to suspicion of all forms of lead. Recent scientific investigations have confirmed what lead miners have known all along: that in the natural mineral form (lead sulfide or galena), the lead is not soluble and is not biologically available, so that health risk from this material is minimal. Indeed, a study carried out by the New Mexico Department of Health in the summer of 1991 where blood samples were obtained from local residents and analyzed for lead showed that there were no elevated levels of lead in the blood of people living nearby the mine site and tailing areas.

However, since scientific studies regarding the health effects of naturally occurring lead were only done recently, state and federal environmental agencies have in many cases not begun to distinguish between forms of lead, and have therefore pressed to apply the same kinds of restrictions on natural lead mineral that they apply to lead paint or some other more biologically available form of lead. This explains why in the summer of 1990, the U. S. Forest Service closed several campgrounds in the Tererro area after they found that mine waste had been used in construction of access roads and other campground facilities. As of this writing, these campgrounds remain closed while the

Forest Service continues to study the problem. Some campgrounds, belonging to the New Mexico Game and Fish Department, were cleaned up in the summer of 1992 and reopened for public use. Meanwhile AMAX began comprehensive study of the mine site to be completed by the end of 1993 with remedial construction planned for 1994 or 1995. It is expected that the remedial work will consist of diverting drainage water away from the pile of tailings, and capping and re-vegitating the the pile itself.

It is interesting to note that as of this writing (mid 1993), AMAX has already expended $3.5 million on the tailings site since 1986, primarily for engineering design, environmental studies , and water monitoring. The reclamation of the tailing ponds, drainage channel, and associated work is to begin in late 1993 and is expected to cost an additional $3.5 million. The completion of studies of the mine site, followed by construction of remedial measures willl probabvly cost another $1.5 to $2.0 million, bringing the overall total to about $9.0 million. This expenditure should result in a much improved appearance of the sites and elimination of whatever health risks, if any, that may be present, as well as minimization of any lasting environmental effects.

This writer, for one, commends AMAX for the remedial actions they are taking, although the costs they are being forced to incur seem far out of proportion to the existing health risks at the sites as compared to to those existing in the millions of homes painted with lead based paint. Howevcer, when the work is complete, the sites will have been truly returned to their former natural state extant before 1926.

ADDENDUM TO EPILOGUE
April 2006

Today as one drives along State Highway 63 in what was once Tererro, the landscape of what had been the mining area has now been totally returned to its naturasl state. The tailings from the mine are now covered with about two feet of topsoil planted with flora of the surrounding region—ponderosa pine, scrub oak, etc. Previous environmentasl concerns of EPA have been totally eliminated. The only remaining vestage of those bygone days is the wooden truss bridge my father and his crew constructed 90 years ago and that is now listed in the National Historic Register. Actual completion of the remedial work was delayed several years due to lawsuits filed agasinst insurance companies by AMAX.

APPENDIX A

TONS OF MINING/MILLING PRODUCTION

Jan 1, 1927 Through May 27, 1939

YEAR	MINING: FROM STOPING	MINING: FROM DEVELOPMENT	MINING: TOTAL	MILLING
1927	156,090	5,939	162,029	164,874
1928	169,796	29,712	199,508	201,013
1929	195,685	21,436	217,121	216,809
1930	128,903	23,207	152,170	152,440
1931	165,874	19,615	185,489	184,502
1932	179,647	5,520	185,167	185,515
1933	186,577	6,085	192,462	191,905
1934	187,861	12,915	200,776	200,839
1935	172,144	12,635	184,779	185,380
1936	145,193	6,081	151,274	150,932
1937	170,405	16,020	186,425	185,850
1938	195,218	7,882	203,100	203,900
1939	70,165	2,792	72,957	75,123
TOTAL	2,123,658	169,924	2,293,582	2,299,082[1]

APPENDIX B

SUMMARY OF PECOS MINES TONNAGE ESTIMATES

JANUARY 1, 1939

LEVEL to LEVEL	NORCROSS ESTIMATE 1925 [2]	ENGINEERING ESTIMATE 1939	MINED to DATE	ESTIMATED BALANCE	AVERAGE GRADE OF MINED ORE: Zn	Pb	Cu	Ag	Au
Above Adit	-------	91,418	91,418	-------	----	----	----	----	----
Adit to Water Level	73,939	179,113	176,113	3,000	14.0	2.75	0.61	1.94	0.06
Water Level to 300	187,145	202,857	199,857	3,000	14.0	2.75	0.61	1.94	0.06
300 to 400	227,848	309,053	308,253	800	14.0	2.75	.061	1.94	0.06
400 to 500	255,205	362,331	360.411	1,920	17.5	5.52	0.57	3.74	0.13
500 to 600	155,999	158,199	125,809	32,390	14.4	4.07	0.41	1.78	0.06
600 to 700	68,540	257,102	254,413	2,689	11.5	2.98	1.32	3.58	0.07
700 to 800	57,120	198,204	191,050	7,154	11.2	4.71	0.45	2.46	0.10
800 to 900	15,000	226,096	221,661	4,435	9.3	4.36	0.44	3.09	0.07
900 to 1000	--------	151,167	126,842	24,325	9.4	3.71	0.54	2.72	0.05
1000 to 1100	--------	80,905	49,640	31,265	9.5	4.09	0.09	4.64	0.09
1100 t0 1200	--------	183,272	129,141	54,131	9.2	3.42	0.19	1.94	0.05
1200 to 1550	--------	47,229	7,824	39,405	9.2	3.52	0.17	0.65	0.03
1550 to 1700	--------	9,067	1,570	7,697	8.4	1.28	0.42	1.10	0.04
TOTALS	1,020,796	2,436,011	2,225,800	212,211[3]	9.8	3.6	0.26	2.20	0.06
KATYDID ORE BODY [4]	309,415	611,609	606,530	5,079					
EVANGELINE ORE BODY[5]	711,381	1,824,402	1,615,270	209,132					

---

1 This milling figure is 5500 tons more than the mining figure. This is accounted for by the fact that 5500 tons of ore was shipped to the mill from development work carried out by Goodrich-Lockhart prior to 1926 (150 tons) and by Amco in 1926 (5350 tons).

2 These figures were presented in Norcross' report to Amco Ltd. management on July 11, 1925 as a basis on which a decision could be made as to whether or not to purchase a controlling interest in the venture. He stated that there might be an additional 500,000 tons.

3 Of this balance, 72,957 tons were mined in the final five months leaving 139,254 unmined tons when the mine ceased operation.

4 These figures are provided to show from which ore bodies the totals shown above were mined.

5 Same comment as 3 above.

APPENDIX C

FINANCIAL SUMMARY

GROSS PROFIT

1927	1928	1929	1930	1931	1932	1933
$733,689.30	$846,689.73	$1,172,254.14	$27,439.20	$23,431.75	$577,833.05	$258,666.39
1934	1935	1936	1937	1938	1939	
$274,204.70	$166,326.25	$15,120.29	$94,778.76	$73,327.65	$55,157.93	

TOTAL GROSS PROFIT $4,318,493.60

VALUE OF METALS SALES

ZINC	$ 9,602,390.44
LEAD	12,708,439.08
COPPER	6,807,714.25
SILVER	5,105,785.24
GOLD	5,440,044.00

TOTAL SALES $39,664,376.01

METAL PRICES

ZINC	APPROX. 6.5¢ PER LB
LEAD	APPROX. 6.4¢ PER LB.
COPPER	APPROX. 13¢ PER LB.
SILVER	BETWEEN 28¢ AND 85¢ PER OZ.
GOLD	$20.67 PER OZ. TO JAN 31, 1934 $35.00 PER OZ. THEREAFTER

APPENDIX D

EMPLOYMENT SUMMARY

YEAR	AVERAGE NUMBER EMPLOYEES	TOTAL MANDAYS WORKED	TOTAL FATAL ACCIDENTS	TOTAL LOST TIME ACCIDENTS	TOTAL PERMANENT DISABILITY	TOTAL PARTIAL DISABILITY
1926	480	*	*	*	*	*
1927	362	128,510	0	**	**	**
1928	389	139,262	6	**	**	**
1929	425	154,445	1	153	1	1
1930	515	187,118	0	108	0	0
1931	605	219,894	1	51	0	1
1932	619	226,624	0	82	0	4
1933	640	231,220	1	84	0	0
1934	741	241,829	2	131	0	6
1935	730	227,708	3	141	0	5
1936	757	259,849	0	174	0	4
1937	859	276,060	1	238	0	7
1938	687	228,848	0	179	0	0
1939	594	76,881	0	7	0	0
TOTAL	600 (AVG.)	2,598,348	15	1,348	1	28

* No Record Maintained During Construction And Development Work.

** No Safety Record Maintained During First Two Years Of Operation.

APPENDIX E

MINE CAMP MAINTENANCE COSTS

	1939[1]	1938	1937	1936	1935	1934	1933	1932	1931	1930	1929	1928	1927
WATER SERVICE	$436.25	963.31	3943.38	3208.17	920.62	876.76	1357.67	2307.27	1706.14	1655.06	1165.39	1051.71	3707.93
LIGHT SERVICE	569.54	506.09	3620.78	5905.45	2790.42	4125.54	2469.62	1976.62	1690.28	2570.14	2997.49	2866.06	3339.01
BUILDING REPAIRS	189.60	870.30	4030.70	4954.18	2212.73	6126.35	1692.30	1390.45	4763.03	3017.59	4289.64	1750.64	6100.42
COMPANY ROADS	381.50	1062.26	2735.19	1695.74	3106.40	1362.54	616.98	809.70	804.62	634.02	2505.40	1393.46	2063.92
SCHOOL OPERATION [2]	1012.45	2263.85	5354.12	2278.96	2131.95	1571.72	2627.45	1807.95	2510.97	870.84	2606.78	2156.24	4146.89
BUNKHOUSE SERVICE	1159.42	2735.74	4639.31	3663.64	4554.64	2894.04	4125.78	3396.18	3919.97	4054.75	10758.15	8220.45	5599.74
DRY HOUSE SERVICE	3745.03	8926.11	11255.31	7948.96	9507.80	6096.07	8146.44	3479.36	4337.22	6467.42	5325.01	4525.11	3330.46
WATCHMEN FIRE PROT [3]	1348.20	3572.92	8557.12	17474.13	3472.57	3095.35	2697.71	2592.21	3004.26	3721.05	3831.57	4750.16	6487.79
CAMP CLEANUP	785.00	1871.77	1610.71	4690.51	1853.09	2005.96	2089.38	1559.58	1329.01	2404.73	1618.09	2504.36	2958.49
STABLE EXPENSE [4]	-----	-----	-----	86.78	59.16	73.50	168.21	85.10	68.07	824.81	1081.22	669.39	452.61
TOTALS		22772.35	45746.62	51906.52	30609.38	28227.83	25991.54	19203.42	24133.57	26220.41	36178.74	30687.58	38187.26

GRAND TOTAL, CAMP MAINTENANCE COST—————$389,492.21

APPENDIX F

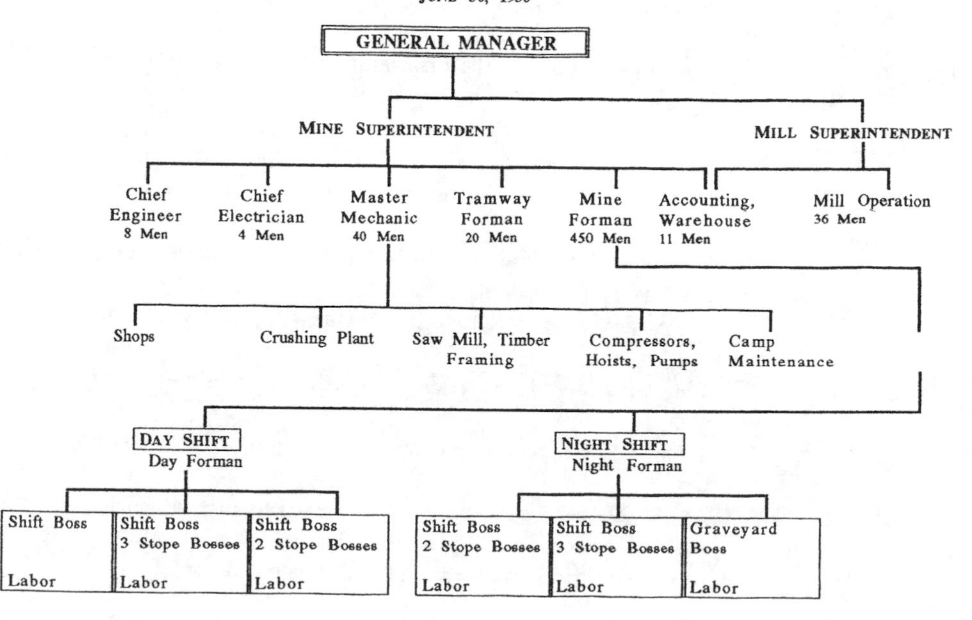

1 January 1 through May 31.
2 Higher costs in 1927, 1928, 1929, 1931, 1933, and 1937 reflect construction of new classrooms.
3 Higher costs in 1935, 1936, and 1937 reflect cost of additional watchmen following strike in 1935.
4 Higher costs in 1927, 1928, 1929, and 1930 reflect need to maintain a stable of horses to carry supplies to tramline and to haul logs for sawmill.

glossary of mining terms

✲ ✲

Adit — A horizontal or nearly horizontal passage driven from the surface for the working or de-watering of a mine.

Apex law — The mining law of 1872 which defines the rights of miners relative to the highest point of a vein of ore, and which gives to the owner of that claim the right to follow the vein along the dip for an indefinite distance without regard to the ownership of the overlying surface.

Assay — The chemical process by which a piece of ore or rock is analyzed to ascertain its mineral content with reference to a given amount of the rock or ore. The result may be provided as a percentage or, in the case of precious metals, as ounces per ton.

Back — The roof or upper part of any underground room or cavity.

Bailing — To remove water from a mine by continuously dipping a bucket, or similar container into the water, hoisting it to the surface and emptying it on the surface.

Barren Schist — A block of schist barren of any mineral ore.

Blasting Round — A number of blasting charges spaced, angled, and drilled to such a depth as to break up rock in a predetermined pattern and shape.

Block — A large piece of detached ore or rock.

Blockholing — The operation of drilling and blasting a detached boulder to make it more easily handled.

Cage — An elevator hoisted up and down the shaft of a mine and used to transport men or material in and out of the mine.

Cap — 1—Pieces of plank or timber placed on top of posts or stulls.
2—The top piece of a three piece timber set.
3—A device for detonating a blasting charge.

Cave-in — The collapse of the back or walls of a mined cavity.

Caved Stope — Type 1—ore is broken by caving induced by undercutting a block of ore.
Type 2—ore is removed by excavating a series of horizontal or inclined slices while the overlying back is allowed to cave and fill the areas previously filled by removed ore.

Change House	A building in which miners can clean up and change from mining into street clothes.
Channel	A six inch wide by one inch deep cut at right angles to a vein. The chips thus removed are assayed for mineral content.
Chute	An underground passage or shaft through which mined material can fall from an upper to a lower level.
Claim	A tract of land staked out by a miner.
Cleat	A small piece of wood nailed to two planks or timbers to hold them together.
Collar	The first wooden framework around the top of a shaft.
Core	A cylindrical shaft of rock obtained by diamond drilling.
Crib	A framework of timbers used to support or strengthen tunnels, shafts, and stopes in a mine.
Cross Bit	A drilling bit with two cutting edges centered on and at right angles to each other.
Crosscut	A small passageway driven at right angles to the main entry to connect to a parallel passageways to either create air circulation or to inspect a specific part of a body of ore.
Crosshead	A tunnel driven at an angle to the dip of a strata to connect different seams or workings.
Cut-and-Fill Stoping	A stoping method in which ore is excavated by successive flat or inclined slices working upward. Ore is removed and the stope is filled with waste up to a level within working distance of the back before the next slice is removed.
Detonator	A device used to explode a blasting charge that is initiated by a safety fuse or an electrical charge.
Development	The work of driving openings to and in proven bodies of ore to prepare them for working.
Diamond Drill	A hollow rock drill that removes a core of rock. Diamonds are embedded in the cutting face of the drill to insure smooth cuts and long life.
Diorite	Any of various granite-textured, crystalline dark rocks containing little quartz but rich in feldspars consisting of sodium and calcium silicates.
Dip	The angle at which a vein is inclined from the horizontal.

Drift	A horizontal or nearly horizontal passageway running through or parallel to a vein of ore, or a secondary passageway connecting two main shafts or tunnels.
Drift Sets	A strong timber set in a drift which may serve as an anchorage for other timber sets in a stope above.
Driven	Having been excavated horizontally or at some inclination as with an adit, drift, raise, or crosscut.
Eductor	A device using the cavitation principle produced in a pipe constriction to pump water. A small amount of water is forced through an orifice in the pipe at high velocity thereby creating the cavitation and causing the water to be pumped to be drawn in and moved to its ultimate destination
Exploit	To mine and market the minerals found in ore deposits.
Extraction Chute	A chute through which material is passed from a stope to a lower level for removal to the surface or another level.
Fault	A fracture along which there has been displacement of the two sides with respect to one another either horizontally or vertically or both.
Flume	An artificial channel or chute for the directed flow of a stream of water.
Footwall	The wall or rock lying under a vein.
Gage	The distance, center to center between a set of rails used for ore cars or other rolling stock.
Gangue	The worthless rock in which valuable minerals are found.
Glory Hole	A funnel shaped excavation on the surface, the bottom of which is connected to a raise driven from an underground haulage level. Waste from the glory hole is used for filling mined stopes or other cavities.
Gob	To store waste rock.
Gob Lines	The underground storage of waste rock along side working places.
Gob Posts	Posts used to hold gobbed waste in place.
Gouges	A layer of soft material, such as talc, along the side of a vein. Gouges favor miners in that they may be removed easily by hand permitting the vein to be worked from the side.
Grizzly	Guard rails made of heavy timber or steel to protect the openings of chutes or manways, or to screen material to some predetermined maximum size.

Guide	The track that supports and guides a skip and/or cage.
Gyratory Crusher	A rock crushing machine consisting of a funnel-shaped steel chamber in which an eccentric head rotates. The rotating head pounds the rock to be crushed against the side of the chamber breaking it into smaller pieces which fall out through the bottom of the chamber when broken to the desired size.
Hanging Wall	A wall that slants inwardly into the working area or stope.
Haulage	The movement of men or material.
Headframe	A framing of timber or steel located above a mine shaft at the top of which is a sheave used to guide the hoisting cable.
Heading	A small excavation driven in advance of the full size excavation. It may also be driven laterally in which event it is known as a crosscut heading.
Hitch Cutter	A tool used to cut a place on rock on which timbers will rest, or the workman who does the cutting.
Inclined Cut-and-Fill	A cut-and-fill operation in which the floor is inclined rather than horizontal.
Jaw Crusher	A rock crushing machine consisting of a set of jaws that breaks rock into smaller pieces.
Lacing	The installation of lagging.
Lagging	The timber or other material placed behind main supports, such as gob posts, to retain the material behind the lagging.
Level	The main underground network of passages located at intervals (usually 100 feet) to provide access to stopes and for the haulage of men and material.
Leyner	A light weight, air driven drilling machine with a pneumatic ram at the opposite end from the drill bit. The ram not only holds the drill in place but provides pressure for the drilling operation. This eliminates the need for manual handling of the drill operation as with jack hammers.
Leyner Drifters	A Leyner type drilling machine used for driving drifts.
Lode	A vein of mineral ore deposited between clearly demarcated nonmetallic layers of rock.
Manway	a passageway, either vertically, inclined or horizontally through which only personnel move from one area to another.
Magma	The molten matter under the earth's crust.

Mineralization	The natural process in which mineral ores were deposited within the earth's upper crust.
Moil	A tool used for manually breaking and wedging out rock or ore.
Mucker	Mine laborers who perform the manual work of removing ore or waste rock.
Mucking	The process of manually loading and moving mined material.
Nipper	One who controls and distributes materials to miners, such as tools or dynamite.
Ore	A natural mineral compound found in the earth's crust.
Outcrop	A portion of an ore body protruding through the soil level.
Patent	A grant made by the government to an individual, conveying to him or her fee simple title to public lands.
Pillar	A column of ore that has not been removed from a stope during mining so as to provide support for the back.
Pillar Mining	A mining technique using pillars to support the roof of a stope rather than cribbing.
Pinching	The predilection of a vein of ore to pinch out and totally diminish.
Placer	A glacial or alluvial deposit of sand or gravel containing eroded particles of valuable minerals.
Pony Sets	A smaller timber set or frame incorporated into the main sets of a haulage level to accommodate chutes, manways, or equipment from above or below.
Rails	The steel members upon which ore cars roll, or which are used in the construction of grizzlies. Size is indicated by the weight per lineal foot of rail.
Raise	A vertical or inclined opening driven upward from one level to connect to a shaft, glory hole, or level above
Riparian water right	A right to use a given amount of water flowing through or immediately adjacent to one's property.
Run	A caving in a mine working.
Schist	A crystalline type rock that can easily be split or cleaved along its parallel layers such a slate.
Second-feet	A measurement of the flow of water passing a fixed point in one second.

Sets — The timber which composes any mine framing whether used in a shaft, a drift, a crosscut, etc. A set may be four pieces constituting a single course in the lining of a shaft, or it may be three pieces constituting a course in a drift, crosscut, etc.

Shaft — A vertical or nearly vertical passage from the surface to levels below, or from one level to others below. The cross section of the shaft is rather limited as compared to its depth.

Shaft collar — The topmost part of a shaft on the surface, sometimes ringed with a collar of timbers or concrete.

Shear Zone — A large scale area of shearing or faulting.

Shoot — A body of ore, generally elongated, extending outward from a vein.

Shrinkage Stoping — Stope mining in successive flat or inclined slices working on the back. After each slice is blasted down, enough ore is drawn off to provide a working space for removal of the next slice above. Generally about 40% is drawn off while the remainder is left in place to help protect against runs in the back or walls. The method can only be used where the rock is strong.

Sill Floor — The floor of any working.

Sinking — The process of driving a shaft.

Skip — A guided steel bucket used for hoisting mined material to a higher level and designed to be easily dumped.

Sludge — The ground up material resulting from the operation of a diamond drill.

Square Set-and-Fill — Successive layers of timber sets for mining higher and higher in a stope.

Stemming — The back-filling of a blasting shot hole, after the blasting material has been emplaced, with sand, clay , or other similar material.

Stope — Area from which ore is extracted in a series of steps.

Stope Blocks — Blocks of ore within a stope.

Stope Filling — The process of gobbing a stope after the ore has been removed.

Strike — The course or bearing of an outcrop of an inclined body of ore.

Stull — Timber props set between walls of a stope or other mining cavity to support the back or roof.

Sub-level — A working level between two main levels.

Sulfite (sulphite) — Compounds of sulfur and one or more minerals. The spelling at that time used the letters p and h rather than f.

Talc — Hydrous magnesium silicate found in the natural state.

Tetryl — An explosive used in blasting caps.

Timber Bearers — The concrete rings upon which lining a shaft are supported.

Toe-cut — A cut in rock obtained by use of blasting charges in singly drilled holes inclined downward.

Underhand Stoping — The mining of a stope by working from its top side downward.

V-cut — A cut in rock obtained by using charges in a series of holes drilled in the shape of a V.
See Fig 7b

Vein — A zone or belt of mineralized rock.

Water Ring — Rings of concrete poured at intervals in a shaft and so designed as to collect any water flowing down the walls of the shaft. The collected water is carried off to a pumping point for removal to the surface. These same rings are used for timber bearers within the shaft.

graphic credits

✲✲

Tererro-Today Sketch	Susan, Huie, and Sherry Ley
Map, Inside Front Cover	Ken Paulsen
Frontispiece, Page ii	Author (Photographer, Osborne C. Wood)
Page xiv	May Walter
Page 83	Author
Page 84	Ralph Littrell, Jr. & Lilah Matson
Page 85	Lilah Matson
Page 86	May Walter & Ralph Littrell, Jr.
Page 87	May Walter & Author
Page 88	May Walter
Page 89	Author
Page 90	Author & Eugene McClellen
Page 91	Ralph Littrell, Jr., Dodie McDuff, & Lilah Matson
Page 92	Lilah Matson & Author
Page 93	Lilah Matson, Dodie McDuff, & Author
Page 94	Virginia DeTevis Francis & Author
Page 95	Author
Page 96	May Walter
Page 98	May Walter
All Other Maps	Author

Printed in the United States
by Baker & Taylor Publisher Services